TECHNICAL MANAGEMENT IN A MANUFACTURING ENVIRONMENT

With Important Tips On Practical Management

BERTRAND WONG

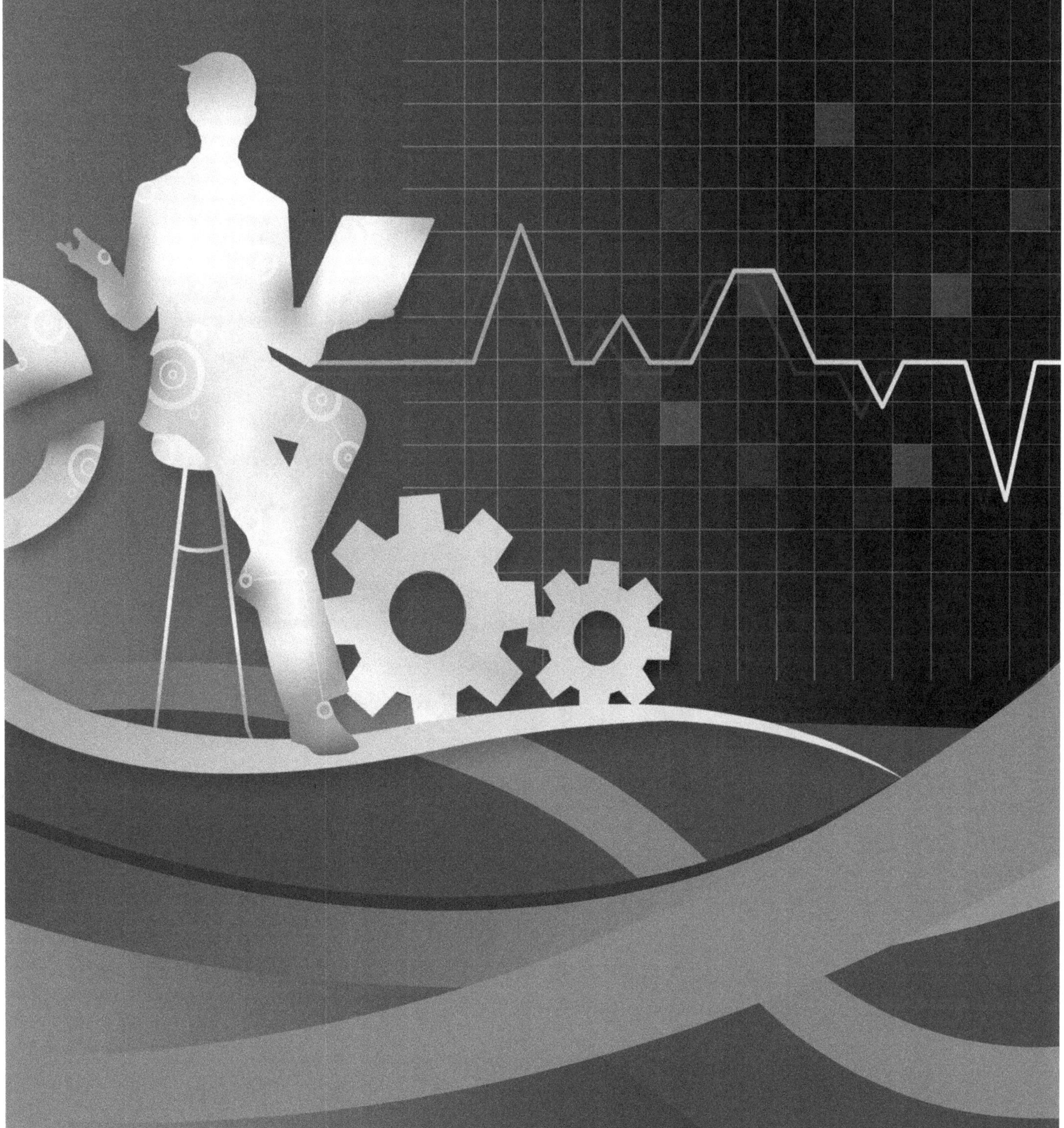

TECHNICAL MANAGEMENT IN A MANUFACTURING ENVIRONMENT

With Important Tips On Practical Management

BERTRAND WONG

Copyright © Bertrand Wong 2012

Important Introduction To Management And Business
Copyright © Raymond Leau 2012

All rights reserved. No part of this publication may be reproduced, stored in
a retrieval system, or transmitted in any form or by any means, electronic,
mechanical, photocopying, recording or otherwise, without the prior permission
of the publishers.

TECHNICAL MANAGEMENT IN A MANUFACTURING ENVIRONMENT
With Important Tips On Practical Management

P R E F A C E

TECHNICAL MANAGEMENT IN A
MANUFACTURING ENVIRONMENT
With Important Tips On Practical Management

To manage and solve technical problems in an industrial or factory environment, book knowledge or theoretical knowledge alone is insufficient, and, practical knowledge or experience is just as important, if not more important. As a matter of fact, some things cannot be learnt from books but must be "experienced" first-hand. This kind of knowledge is of great importance to the practising engineer or technical manager.

In this book the author shares this kind of knowledge which he has gained first-hand by being "on-the-job" with the reader, knowledge which is unlikely to be found in a normal technical management book. There are also important tips on practical management.

Bertrand Wong
Ph.D. (Engrg.), Ph.D. (Bus.),
PE, Certified Professional Manager,
Vice-Chancellor, Research Professor, Eurotech

CONTENTS

1. Important Introduction To Management And Business
   By Raymond Leau, Managing Director,
   Prudent Properties Pte. Ltd. 1

2. Introduction To Technical Management 3

3. Case Studies/Examples On Various Aspects Of Technical Management

   a) Module 1: Production Engineering 5

   b) Module 2: Design Engineering 15

   c) Module 3: Quality Engineering 23

   d) Module 4: Sourcing Of Parts 28

   e) Module 5: Sales Engineering 29

   f) Module 6: Technical Communication 36

   g) Module 7: Maintenance And
      Calibration Of Factory Equipment 39

4. The Technical Manager And Modern Technologies 42

5. The Latest In Electronic Manufacturing 44

6. Conclusion 47

7. Important Tips On Practical Management 48

Appendices: Exhibits Of Engineering And Other Important Documents 51

IMPORTANT INTRODUCTION TO MANAGEMENT AND BUSINESS

By Raymond Leau, Managing Director, Prudent Properties Pte. Ltd.

Managing technical people seems to be another kind of "ball-game". Many technical people are proud of their expertise and would not like to be always told what to do, as though they do not know their stuffs. Moreover, technical knowledge is advancing so fast that the current technical knowledge could become obsolete within a short time. This book provides many important ideas for the aspiring technical manager. Much of the knowledge presented in the book is not available in textbooks and could only be gained by being on the job.

However, as a person with years of experience in management and business, I would like to share some of my experiences with our dear readers.

First of all, business is a venture fraught with all kinds of risks, especially the risks of export business, e.g., fire hazards, industrial accidents, recessions, financial losses, political instability, even war, etc.. Not everyone is willing to take such risks. As they say, businessmen do need to take calculated risks. But how do we calculate the risks? What are the risks which are acceptable? Here experience, judgement and intuition have to play their part. I would say that in business, we win some and lose some. It is impossible to win all the time. We just need to minimise the risks, which are unavoidable.

We should be in a business for which we have experience in and are confident of success. Otherwise, we should engage someone trustworthy who has the necessary experience and ability to run the business for us.

Next, we should invest only to the extent the amount of money our company is prepared to lose should the business go belly-up, and, have a contingency plan to back this situation up, to overcome this adverse situation. Many business people do not plan for contingencies and end up in hot soup when they get loans from loan-sharks and lose further sums of money, making loan repayments practically impossible.

Then, we need to recruit and train the right people to do a job well. It is pointless to recruit somebody to perform a function just because he or she is cheap. Has he or she really the aptitude, ability or interest for the job? We should have our priorities right. It would be problematic if our organisation has a capable manager who is a crook who cheats on the company, which could cause the company to collapse eventually. You see, to be a good crook, a person also needs to be capable. We must have people who use their capabilities to work for us and not against us. So, recruitment of the right people should be a patient, cautious and thorough process. Therefore, honesty and integrity should be given a higher priority than capability. In fact, I think they are the most important for an organisation to survive and progress.

The other thing is that we should not encourage and should discourage office politics in our organisation. Many organisations, especially the large ones, are full of cliques and office politics. When this happens there would not be real team-work. These office politicians might appear to work together as a team, but it is all only acting, and, we know that these so-called office colleagues are just simply watching out for themselves, being suspicious of each other. As a manager, we should be able to judge and see who these office politicians are and should not fall for their flatteries and tricks. In short, we should all discourage cliques and office politics and encourage real team-work.

As you know, the business world is a fast-changing place. Technology appears to be changing faster than we could catch up. Market conditions are in constant flux. When a business does well, many would join the industry, resulting in price undercutting, even price wars, staff poaching, and other competitive activities. Many companies have set up branches outside Singapore, as the market here is too small and too competitive. They just go elsewhere to look for greener pastures. If they do not venture overseas, the business here is so reduced and small that they might have to close shop. However, venturing overseas, where the culture is different and unfamiliar, where there might be political instability, is even more risky. Unless we are able to accept such risks, and calculated ones too, we cannot think of venturing overseas. And, we would have to close shop if the shrinking local market cannot sustain the business.

We have to also bear in mind that many products in the market have a life cycle, which is: product introduction → growth → maturity → saturation → decline. When a product reaches the "decline" stage, we have to be prepared to drop the product. Then, we should introduce a new product or products with market potential as a replacement. However, many business people do not seem to understand, or are even aware of, this market condition.

I would also like to share my experiences in managing people. It is never easy to manage people. I have had my share of staff problems, just like any other manager. I have had staff who left our company because our competitors pay them more. Some might have left because they did not like to work for me. Nowadays, people are more demanding, especially in Singapore, which has become a developed country and where the standard of living and educational level of the people are very high, resulting naturally in higher expectations. It is nowadays difficult to find loyal, dedicated staff. If we find staff who are loyal, dedicated and hardworking, we are indeed very lucky and have to try our best to take care of their welfare, give them challenging, satisfying assignments, even promote them, and, of course, retain them. As we all know everyone has needs and their characters and interests might be very different. It is not going to be easy to manage a group of people with diverse personalities, interests and motivations. Some, especially Singaporeans, do not like to and cannot work night-shifts or weekends, while others have no problem with that. It is of course very important to find dedicated, hardworking and responsible people who do not mind working in not so ideal working conditions such as night-shifts and weekend work. It is actually easier said than done. Such ideal staff are also not readily available, resulting in many local companies employing foreigners who do not have so high expectations where salary and working conditions are concerned.

One solution for an organisation pertaining to such manpower problems is of course to automate as many of the company operations as possible and rely as little as possible on human labour. But certain jobs such as those of selling cannot be automated. The other way out is to switch over to a business which relies less on human labour.

Management however is an art, not a science. We have to be prepared for changes, to learn from our past experiences and mistakes, and to make improvements, to change for the better. This book provides plenty of practicable ideas for the aspiring technical manager to tap on. I hope that technical managers would be able to benefit much from these ideas when implementing them.

INTRODUCTION TO TECHNICAL MANAGEMENT

Managing technical personnel is hardly an easy task. Technical personnel by virtue of their unique expertise or experience have a tendency to resent interference and too much supervision; as far as they feel that they are more knowledgeable than their technical manager, they are likely to resent the latter's control.

Nevertheless, technical management is a manageable task that requires some skill and experience. Experience is especially important as it gives confidence.

In this book the author shares his experience with the reader, especially the reader who is lacking experience in technical management. It is hoped that the reader, when reading the book, would have a feeling of "being on the job itself".

The author now brings the reader back to his former place of employment, a multi-national consumer electronic manufacturer, where he had been engaged in technical management. In the following chapter, he presents a diary of engineering events/meetings to acquaint the reader with the nitty-gritty of technical management.

But before we proceed to the next chapter, it is perhaps necessary to enumerate the functions of the engineering department. As the person responsible for the engineering department, the author had been overseeing the following functions:-

1) Troubleshooting and finding solutions to technical problems.

2) Product design and performance improvement.

3) Cost saving, productivity improvement, quality improvement and value engineering.

4) Pre-production of new models.

5) Preparation of samples or sample models for clients.

6) Settlement of problems of quality posed by both production and quality control departments (especially with regards to disputes).

7) Work study and methods improvement, including establishing time standards (which are required for production planning).

8) Design and preparation of jigs and/or fixtures (especially for the introduction of new models into the production floor).

9) Sales engineering, i.e., liaison with clients with regards specifications and technical problems, including entertaining them.

10) Calibration and maintenance of test equipment and instruments.

11) Incoming quality control, i.e., check on quality of incoming parts, components and other items.

12) Oversee introduction of new models into production line and see that it is carried out smoothly without any technical hitch.

13) Carrying out of drop tests, and temperature and moisture tests, after products are packaged to ensure that packaging provides sufficient protection for products while the latter are in transit containers, etc..

14) Installation of new plant and equipment.

15) The aesthetic aspects of existing and new product models, e.g., colour, texture and shape.

16) Engineering change control, e.g., issuing and changing part numbers, including amendments to engineering drawings.

17) Preparation and issuing of engineering drawings to production, quality control and materials departments, with regards to new models and changes in product or electronic circuit designs.

18) Preparation of technical information reports for internal department circulation and or circulation to other departments.

19) Preparation of technical manuals for internal use and reference, i.e., ensure that technical knowledge or know-how is passed on and shared.

20) Liaison with sub-assembly contractors. Check on their quality together with quality control department, visit their plant to check product quality when they are carrying out sub-assembling of new models.

21) Preparation of parts comparison lists (comparison of parts of old model and parts of new model).

22) Supply all necessary engineering drawings and technical information to production and quality control departments with regard to new product models prior to their introduction into the production line. Explain new models to production personnel concerned to ensure that they know how new models are to be produced, especially seeing that they fully understand the clients' specifications. The same goes for the quality control personnel concerned too, so that they would know what to look for when inspecting the new models. Supply them clients' specifications, of course.

23) Test engineering, i.e., testing of models by using instruments such as digital readers and voltmeters.

24) Liaison with technical experts from head office.

25) Research and development activities.

26) Preparation of user's handbooks.

27) Machining work and fabrication of parts.

28) Sales engineering, i.e., exhibiting and explaining product(s) to clients and potential clients, including attending to clients' complaints/problems.

The following is an organisation chart of the engineering department:-

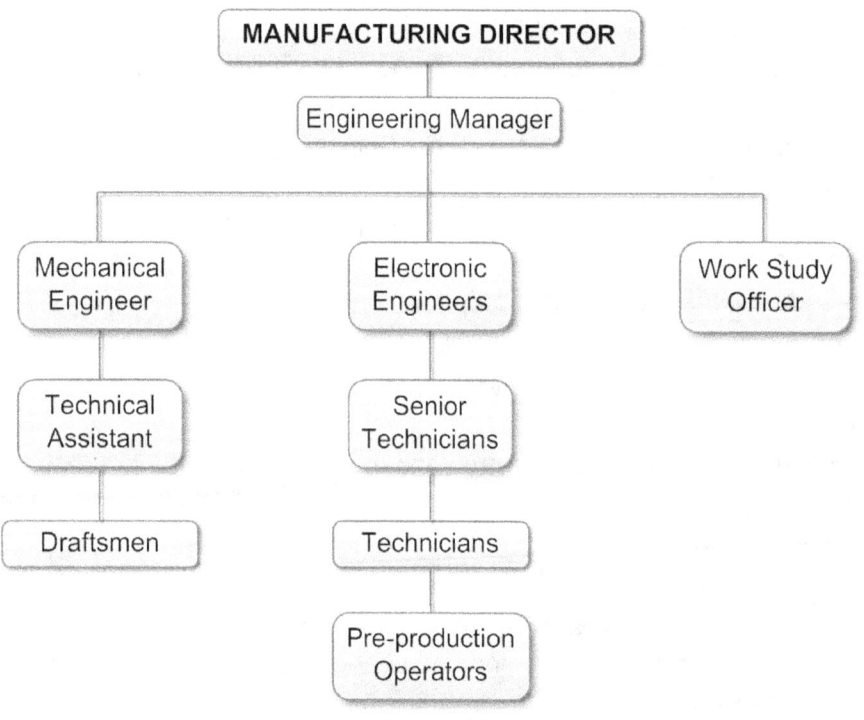

CASE STUDIES/EXAMPLES ON VARIOUS ASPECTS OF TECHNICAL MANAGEMENT

a) MODULE 1: PRODUCTION ENGINEERING

Case Example On Production Scheduling And Production Problems

1) L Model
2) Re-schedule production
3) UK/Australia version
4) RHOI Resistor for Model 2000 to be 21 kilo-ohms, instead of 8 kilo-ohms.
5) R 41 Resistor for Philippine version, Model 1000 PH to be 22 kilo-ohms instead of 8.2 kilo-ohms. Less than 2,000 sets have been sent to Australia since last year.
6) Production schedule for Australian models is as follows:

	Model 2500 P31	Model 2000 L	Model 2000 US	Model 2500
May	150 sets	200 sets	100 sets	100 sets
June/July	Nil	Nil	Nil	Nil
August	150 sets	200 sets	100 sets	100 sets

7) Production schedule for UK Model TS 2500 WU is as follows:

May	: 250 sets
June/July	: Nil
August	: 250 sets

8) 300 cabinets for model to be molded by 8th. May.

Case Example On Production Management

1) Panels for Models 2000 (GM-12) and 2500L (GM-12T) are to be darker (front and back) for the June and July lots.
2) 14" colour TV to:
   a) Be free-scanning
   b) Have analogue input
   c) Operate within the 15 Khz to 22 Khz range
   d) Magnetic field of CRT to be 0.3 gauze
3) To develop CPU.
4) Inventory for Model 1000 (for Hong Kong) is to be reduced. Lead time is six months.
5) Model 2000 HK to be changed to 2000 US.
6) PQR Co. confirmed order for 1,000 units of 2000 US for August delivery.
7) Monitor to contain enhanced graphic adaptors (EGAs).
8) Memory storage capacity of CPU is 128 Kbytes.
9) Also necessary to include graphics memory expansion card and to obtain graphics memory module kit.
10) Panel of Model GE 8017 to be light grey or natural black in colour.

11) Gamma test is to be carried out.

12) Delivery of 14" sets to Microbee. (Brand-name?)

13) Alpha-Beta is to carry out hot-stamping and to provide sample.

14) Label is to be matrix label. Head office would supply sample.

15) Label material sample, including ink sample, to be submitted, as well as label layout samples. The ink has to be chemically tested.

16) Bar code labels on carton boxes to be pasted on one side of the box.

17) Samples would be submitted to Alpha-Beta, two samples by 21st. May.

18) Carton box sample ready by end of May.

19) Two samples to be reprinted - outside to be white in colour. To confirm printing colour via Alpha-Beta.

20) AC power cord and signal cable.

21) Power plug to be right-angle type (Maker is Belden. Country X distributor is Intertech). Plug, which is moulded type, would be supplied together with cable.

22) Colour of power plug and signal cable. To contact Alpha-Beta or Bravo-Charlie, which are supplying cables for PQR Co. products in Country X. Letter of authorisation already given by PQR Co..

23) Delivery improvement schedule - schedule to be shortened by 18 days - this entails working overtime and in summer holiday. But PQR Co. has to pay extra $58,367.10. PQR Co. to confirm by 25th. May.

24) Confirmation of first lot of monochrome monitors. Shipment to be by air or by sea - by partial shipment. We would be informed about mode of shipment by 20th. August.

25) PQR Co. would send forecast for the six-month period commencing from October this week.

26) AQL submitted by Alpha-Beta approved by PQR Co. (major: 1% & minor: 2% (2.5%)).

27) EMI test for PQR Co.. Liaison with Mr. Donald. Shield for colour graphic card for EMI measurement.

28) Cost of paper white CRT is up by US$30, which PQR Co. cannot accept.

29) Submission of UL/CSA critical component list by 20th. May.

30) PQR Co. sample for RS (European version). Quotation for 5K,10K, and 25K, by 18th. May.

31) Samples - 13 sets by mid-July. PQR Co. wants to know when we can submit final VDE sample.

32) UL application by multiple listing.

33) Test equipment - 1 disk drive from PQR Co. brought back.

34) Source inspection - to be confirmed by PQR Singapore.

35) Problem - Alpha-Beta & Foxtrot-Gopher staff cannot speak English.

36) Specifications deviation - write formal letter to PQR Co.. PQR Co. must write a deviation notice - change from 35 watts to 55 watts.

37) 98 sample sets to be sent by 28th. June.

Case Example On Production Management

1) 12" Model T-125 to be paper white.

2) To have dual frequencies and graphic capabilities. Vertical and horizontal frequencies to be 15.75/18.75 Mhz and 22/21.85 Mhz.

3) To have enhanced graphic adaptor (EGA) and 664 gray scales.

4) Perhaps, dual frequencies of 15.75/22 and 15.85/33 are better.

5) To have auto-scan as well.

6) CTV Model U-3 to have voltage of 120 volts (US version).

7) Model 1404 to be colour.

8) To be PAL (Phase Alternating Line) system.

9) PQR Co. problem - Intertech cable.

10) Contract to be signed between Zerox and Mr. Pearson.

11) Mechanical matters - tooling cost of two links is US$300,000. Requesting additional US$2,400 plus US$47,795, giving total of US$350,195, which is pending approval from head office.

12) Colour of cabinet is to be pink.

Case Example On Production Scheduling

1) To keep stock of some U.K. and Australian models.

2) Models to be produced end of May.

3) Schedule for models as follows:-

	MAY	JUNE	JULY
1000HK		660	660
1000US	660	660	
2000L	200	200	200
2500LS	100	100	100

4) Stock-taking for SA 1000.

5) 150 units of 1000 SP to be immediately taken by XYZ Co..

6) All SP models are similar.

7) Ten sets of 1500 US to be delivered by 10th. June.

8) There is capacitor problem for following models:
   a) 2000 Australia (2/3 months' stock required)
   b) 2000 UK
   c) 2000 Australia

9) Australia to inform us when to send all stocks from our warehouse.

10) Prices for Thailand and India.

11) Two samples for OPQ Co.. Power to come from computer. To be DC (Direct Current) operation. Synchronisation signal and power supply to be same as Australia.

12) Cost of STU Co.'s moulding die is US$33,600 FOB UK.

Case Example On How Mass Production Is Carried Out

1) Mass production for Model 1954.

2) Get ready electrical and mechanical specifications.

3) Check production.

4) What are the problems of production?

5) Explanation of the procedures involved.

6) Pre-production of the sets.

7) 10th. June activities are to be as follows:
   a) Mr. Stephenson to prepare test-com jig for Model 1954.
   b) He is also to run signal cable from signal room.
   c) He is to instal test equipment.

TECHNICAL MANAGEMENT IN A MANUFACTURING ENVIRONMENT

8) Model 1297 has ringing problem.

9) New conveyor is to be ready by 10th. June.

10) First lot of Model 1298 is to be shipped by mid-July.

11) Altogether, 2,000 sets of 2000 Australia, 2000 UK and 1297 are having ringing and overshoot problems. The cause of the problems is resistor R401.

12) A flow-chart is to be prepared, showing countermeasures for Model 1000 US.

13) Moulds for Alpha Co. are not urgent.

Case Example On Preparation For Mass Production

1) Signals cable and programmable generator to be tested.

2) Mr. Lim to prepare gages.

3) Mr.Toh to prepare offset gage and box generator for Model 1297.

4) CPU for Model SP 2000 UK/Australia to be checked.

5) Preparation of jigs before production starts.

6) Humidity test.

7) To recruit work study officer.

8) QC to check Model 1954.

Case Example On Introduction Of New Models

1) Pre-production of 1942 and 1944 to commence any time now.

2) Engineering staff to prepare jigs and introduce models to line leaders and supervisors.

3) Time standards to be ready one week before mass production on 23rd. August.

Case Example On Preparation For Mass Production

1) RS chasis for RA 1342 U and PA 1344 V.

2) Note is to be taken on change in following parts:-
   i) Resistors
      a) R116 (820n - 1.8Kn)
      b) R115 (390n - 2.2Kn)
      c) R113 (2.7Kn - 4.7Kn)
   ii) Capacitor (Pending Approval)
      C301 (10uF 50V → 0.47uF 50V)

3) Reason for change - To decrease current flow to IC 101 and improve buzzle on sound output.

4) Problem of CA 1342 R and CA 1344 R and improvement:
   a) Picture cannot cut-off completely. Cut-off level is around 85% to 90% (After cut-off luminance is 0.6 mite).
   b) Picture quality is normal though.
   c) After change in parts, sound output buzzing improved and there is picture cut-off.

4) Problem of 33DDII and improvement:
   a) Picture can only cut off completely when brightness and contrast are set at minimum.
   b) Picture quality has been normal all along.
   c) After change in parts, sound output buzzing improved and there is picture cut-off.

5) Moulding die ready by 12th. August.

6) Back-covers are moulded in Singapore.

7) Front panels ready by end of August.

8) Pre-production of PL 1435 front panels on Monday, Tuesday and Wednesday.

9) Locally injected front panels, 54,000 units of them arriving tomorrow.

10) One set each of "preproduction" sets to be sent to Johnston-Kingston Works.

11) 13 sets (VDE sample) of SA 1260Z to be sent to XYZ Co. by mid August.

12) 98 sets of SA 1260Z to be sent to XYZ Co. by 28th. August.

13) Mass production of SA 1260Z to commence by 31st. August.

14) Colour samples of CR 1329 monitor to be sent to PQR Co., Australia, by mid-September.

15) The following are to be ready by 23rd. August for new production line:-
   a) Electronic test-com jig
   b) Signal cable from signal room
   c) Installation of test equipment

16) New production line to produce 500 TV sets per day.

17) To prepare parts difference list for Models SQ 1299X and SP 2000.

19) Parts differences between CA 1416 PS and CA 1416 PCM are as follows:-
   i) Holder terminal: 440 C 06702 - 440 C 12703
   ii) Holder - Anterior - Omitted and Screws Omitted
   iii) Band - km: one piece - two pieces
   iv) Screws: two pieces - three pieces (for mounting anterior terminal to back-cover)

20) Procedures for introducing new model as follows:
   a) Electrical measurement data
   b) New model registration/product history record
   c) Performance check record
   d) Product manual, tentative specifications and photographs
   e) General specifications and photographs
   f) Confirmation of all printing matters
   g) New design specifications information
   h) Safety approval application (to safety board)
   i) Pre-production report

20) Procedures for mass production as follows:
   a) Pre-production (sample sets)
   b) Preparation of jigs (to be ready before mass production commences)
   c) Vibration and drop test with sample set CS
   d) Performance check on sample set CS
   e) Prepare time standards (for production planning use)
   f) Introduction of model by engineering staff to personnel of production and QC departments
   g) Mass production

21) Two sets each of D 3416 and D 3417 (both having 14" Hitachi tubes) for test - one set each urgently needed for signal test.

22) Line flow charts for production of these models are needed.

23) Pre-production of five sets of each model.

Case Example On Introduction Of New Models

1) Parts for CR 2034 LQ to arrive between mid-September to end September, which is to be followed by pre-production.

2) Parts list to be ready in next few days.

3) Purchase order for equipment by mid-August. Equipment to arrive by mid-October.

4) To obtain five kits for model from Johnston-Kingston Works.

5) Jigs are OK.

6) VHF (Very High Frequency) & UHF (Ultra High Frequency).

7) Switch generator to be ready.

8) Intermediate frequency carrier.

9) NTSC-type tube. Senior technician, Mr. BC, to prepare.

10) 1434 B to be on PAL system.

11) 1305 A to be on NTSC.

12) Fit 1304/1305 A CRT (cathode ray tube) into 1434 panel.

13) Parts must arrive by 25th. September.

14) Pre-production for E1416 and E1417.

15) Plastic cabinets for CR 2034 LQ.

16) Lima-Mina to make 300 sets of 2034 LQ by early September.

17) Subcontractor, STU Co., to make 400 sets.

18) Mass production of CR 2034 LQ to commence in early October.

19) Colour monitor, Model 9 P 1405 C, DEF Co. version, to be mass produced in early December.

20) To confirm prices of local parts.

Case Example On Production Management

1) On 25th. August, new production line to produce CTV, Model 1864 R, 5,000 units.

2) After that, new model, 1854 R, to be produced.

3) Sister company to produce chasis for 1,000 sets.

4) Slightly more than 200 sets to undergo aging. Each set to be "aged" for 1 1/2 hours. For aging, there are four racks, each rack having three shelves, each shelf capable of having 11 sets. This means that the sets are moved from the shelves about 5.6 times per day. 528 sets can be "aged" per day.

5) Lines 1 and 2 to be under the new production supervisor.

6) Samples for aging to be taken from Line 3.

7) Mr. QRS to be in charge of Line 4.

8) 1864R/1854R panels to be ready by 10th. September.

9) 100 Units of SU 1260 Z as sample.

10) To introduce colour samples of SW 1298 X by 20th. September.

11) Front panels of SW 1298 X, which are without texture, are ready by next week.

12) Mass production of E 1416 to be in mid-December. Front panels by BCD Co. are ready by 14th. September.

13) Mass production of E 1417 to be in early November.

14) Drawing for E 1416 to be issued by 25th. October.

15) Drawings for E 1417 to be issued by 10th. September.

16) Panels of SW 1298 X to be textured in Country X. Mass production of SW 1298 X to be postponed to 20th. December.

17) Checking of jigs for 1864 R and 1854 R.

18) Key-boards, pre-amplifiers and NTSC signal generator to be ready.

19) Subcontractor, STU Co., to start producing by 26th. September.

20) Good Manufacturing Practice audit on 9th. September at 10 am. To be prepared.

Case Example On Preparation For Mass Production

1) Have all parts for SU 1260 Z arrived?

2) Are jigs for SW 1298 X ready?

3) Schedules for SU 1260 Z and SW 1298 X.

4) How about signal room and other equipment?

5) What equipment are needed for E 1417 and E 1416?

6) What equipment are needed for colour monitors?

7) Flight details for one set of CQ 2034 S, two sets of E 1417 and one set of E 1416 as follows:-

IV No   : K 95103
HAWB No.   : MSAS - 473116
MAWB No.   : 160-427/5691
Flt No.   : CX 503/711/28
ETA Spore   : 28th. September

Case Example On Introduction Of New Models

1) Introduction of new models.

2) Value engineering - all unnecessary parts and components are to be taken out.

3) Check production line adjusting jigs and arrange for their storage space.

4) About 320 operators are needed.

5) There are presently four long "aging" racks each giving accommodation for three layers (11 sets per layer), making it possible to "age" 132 sets at any one time.

6) The factory has "aging" capacity of 769 sets per day.

7) To carry out the following checks on the chasis of Model 1954:
   i)   Detail check
   ii)   Appearance check
   iii) Performance check
   iv)   Alignment performance check

8) Get "limit focus" sample of 1954 from outgoing Q.C. and check whether there are vertical jitters, top curl and faulty dynamic focus.

9) ABC Co. to order 10,000 sets of monitor each year. Price per set will be between $185 and $199.

10) XD 1405 D (for ABC Co.) to have the same standard as our company brand's.

11) ABC Co. will initially order 300 sets, the actual quantity to be ordered depending on the price.

12) The market for the sets will be South-East-Asia (Singapore, Indonesia, Hong Kong, Philippines, Malaysia, and Thailand and, also, Europe (Australia, West Germany, Belgium, Holland, Spain and Italy).

13) Specification confirmation:

    a) Packing case to be black in colour - same as ABC Co. "EGA" model - EGA (Enhanced Graphic Adaptor) to be changed

    b) Name plate to be based on our company brand's

    c) "Made In Country X" label to be omitted

    d) Cabinet colour to be IBM ivory

    e) Logo "ABC" to be at centre of panel

    f) IB to be based on our company brand's IB

    g) Power cable to be direct connecting

    h) To reduce cost

    i) To obtain parts locally

    j) Specifications to be re-checked

    k) Cabinet material to be evaluated

    l) One set of monitor needs to be sent for evaluation

    m) The following schedules are to be adhered to:

        i) 10th. November - ABC Co. issues Purchase Order to our company (after final specification confirmation)

        ii) End November - our company issues Purchase Order to head office

        iii) Beginning to mid-February `89 - Pre-production of ten sets

        iv) March `89 - Arrival of parts to our factory

        v) Mid-April `89 - Commencement of delivery of sets

        vi) Mid-May `89 - 300 sets to arrive at ABC Co.

        vii) If new printing die were to be made in Country X, we have to find out the cost

14) IC indications are as follows:

    a) Linear ICs

    b) (i) Bipolar, Memory, TTL, or ECL ICs

       (ii) Memory, RAM, or ROM (Microprocessor) ICs

    c) Hybrid ICs, e.g.:

        i) IC MN 6178 FVE - Linear Audio Power

        ii) IC 14 DN 201 - CMOS - 64K RAM

        iii) IC MN 15847 FVL - NMOS - ROM

        iv) IC M74LS74AP - Bipolar Ringer

        v) IC BX 6221 - Microprocessor

        vi) IC Hybrid STK 4017 - Audio Power

       The above ICs are available at M & M Ltd.

15) Two of our technicians are to check three units of 480P646AI speaker for Model 2103 D (N) on following:-

    i) Sound output level

    ii) Sound pressure

16) Circuit change to be according to following:-

FC CONNECTOR

1) Original Condition	2) Circuit Change	3) Circuit Change + SP Change →
Use same condition Sound output (W)	Centre (Max)	
Sound Pressure (Actual)		

17) The following communication is given to our American Office by our Alpha-Beta Office:-

   i)   ABC Co. inspector to visit Country X, ABC Co. may want to change schedule.

   ii) Has ABC Co. agreed to change of cabinet materials?

   iii) Are there any news on ABC Co.'s next order?

   iv) Alpha-Beta has sent colour samples of cabinet to ABC Co., New York, on 4th. October. ABC Co.'s (New York office) approval is needed as soon as possible. Our American Office to follow up on this matter.

   v)   ABC Co. has accepted carton box sample despatched from our factory though it is a bit different. (This matter should have been discussed when Jim from our American Office was in Country X.)

Case Example On Production Planning

1)   UL (three sets) and CSA (two sets) application for TW 1270 Z on 15th. August.

2)   November - 1,000 sets of TW 1270 Z to be produced.

3)   To order signal cables, sliders and pads.

4)   Use Tandon CPU for tests.

5)   Use IBM video card.

6)   Change IC (location U37) to PQR Co. EPROM.

7)   Schedules as follows:-

   i)   1288 Z - Mass production on 25th. October

   ii) 1942 - 1,500 sets to be produced on 25th. October and 2,500 on 11th. November

Case Example On Factory Management

1)   Follow-up on Good Manufacturing Practice audit:-

   i)   Internal calibration of production instruments and equipment (Engineering Dept.)

   ii) Work-in-process assembly unit in Chasis Section should be clearly defined (Production Dept.)

   iii) More detailed calibration measurement data should be provided by Mitutoyo Calibration Centre (Outgoing Q.C. Section)

   iv) Resistance measurement data of Hi-Pot resistor box should be kept as records (Q.C.,Production and Engineering Departments)

2)   Internal problems/matters:-

   i)   Cleanliness

   ii) Flow of information

   iii) Technical (Design, production, quality control, purchasing, etc.)

   iv) Inter-departmental relationships

   v)   Location of equipment or tools

   vi) Scheduling of jobs

vii)   Safety

viii) ECN system

ix)   Communication with Country X

x)   Wastage (Stationery, electricity, etc.)

xi)   Engineering staff relationships

Case Example On Production Management

1) Production of 35 CC 22 to commence soon.

2) No. 1 Line has started producing CP-1345 last week. Has finished around 222 sets. Production is two days behind schedule.

3) How many PCBs were made at our sister company? Purchasing to liaise with sister company, and check quantity for tomorrow's production. (At least one day's stock ahead of production is required.)

4) Ask sister company to give their production schedule.

5) Sister company revised schedule for CP 1344 S.

6) 1954 is on schedule.

7) For CP 1344 S our factory is behind schedule by 1.5 days. (Can do overtime if PCBs from sister company come in.)

8) Check with sister company everyday.

9) 1942 is about 1,100 units behind schedule.

10) PG 2000 US - Power cord problem? Power cord is not listed. To have green label.

11) Warning label to include the words, "And The Same Rating Fuse".

12) Consider the following carefully:

   a)   Distortion

   b)   Brightness

   c)   AC Power Cord strength relief test.

   d)   Warning Label

   e)   Turning/Bevel of monitor

13) Production schedule for F 3014 T and F 3015 T as follows:

   a)   <u>F 3014 T</u>

        April `87 - 500 units

   b)   <u>F 3015 T</u>   - With Hitachi Tube

        March `87 - 1,000 units

        April `87 - 1,250 units

14) Arrange parts difference lists for F 3014 T and F 3015 T.

15) Prepare jigs.

16) Colour monitor line to be in operation from April `87 onwards.

17) We have time till 1987 to modify existing monitor line for colour monitor production.

18) There is an increase in the price of semi-knocked-down (SKD) sets.

b) MODULE 2: DESIGN ENGINEERING

Case Example On Model Improvement

1) Should prevent mistakes of Model 202 in Country X.
2) Engineer from Hytech to be sent here.

Case Example On Design Modification

1) July - to produce 200 units of model SP-0901 XU.
2) Improve design of monitor circuit.
3) Comparison with Prestige's monitor.
4) Feel that our monochrome monitor is not good enough.
5) First shipment to be made in August.
6) 23rd. April `86 - AMA people from Taiwan to check production facilities.
7) Model 1490 - problem with Diode TRR. Recovery time is longer.
8) To check cathode ray tube and cable.
9) Model 2000 PS for XYZ Co..

Case Example On Management Of Technical Problems

1) Two samples of bar code label to be sent to Alpha-Beta by 15th. July.
2) Carton box sample to be sent to Alpha-Beta by end of July.
3) Extra $58,000 to be informed by Positron Co. by 21st. July (shortened by 18 days).
4) Forecast of delivery for six months from last week of October to be prepared.
5) UL/CSA critical component list is to be submitted by Positron Co. by 20th. July.
6) Positron Co. sample is to be ready - to submit quotation for 5,000, 10,000 and 25,000 units by 2nd. August.
7) 13 sample sets to be sent to Positron Co. by mid-August.
8) When can we submit final VDE sample?
9) 98 sets of colour monitor to be shipped to Positron Co. by 28th. August.
10) Colour monitors to have dual frequency, video input with data grey tube and enhanced graphic adaptor (EGA), and to be IBM compatible.
11) Mr. Hudson to do costing of the 98 sets.
12) The 98 sets are to be tested for 10,000 hours (aging). 20 sets are ready now.
13) Ratings to be as follows, to prevent overheating:
    a) Capacitor Voltage < 0.5
    b) Resistor - Wattage < 0.5
    c) Coil - Temperature < 0.5
    d) Transistor - TCR < 0.5
14) Time standards for Model CS 1954 R to be prepared.
15) Diskette to be provided by Positron Co..
16) Manpower problem. Who will test monitors? Will QC be able to provide sufficient manpower for this task?

Case Example On Technical Problems

1) The following are results of drop test on Model CQ 1954 R (dropping the TV and monitor sets after packing in carton boxes and checking the effects of dropping, to see whether the sets could withstand rough handling by delivery/cargo agents):

   a) Cathode ray tube (CRT) shifted 3mm down

   b) Printed wire board (PWB) shaky

   c) CRT guard cracked

   d) Stress strides on nut

   e) Variable resistor (VR) not functioning

2) How can CQ 1954 R's problems be prevented from occurring again?

3) Production of CQ 2102 P, SQ 1298 and CQ 2102.

4) Chasis from sister company to be used in preproduction of CQ 1954 R.

5) Rough schedule for CQ 1954 R as follows:

   a) 10th. August - Introduction and preparation

   b) Four days before mass production - To check torque of CRTs, and change nuts to hexagonal type, and screws to another type

6) Resistors R 115, R 116 and R 113 to be substituted by R 555.

7) Pre-production of 1944 and 1942 to be carried out 1.5 months before mass production. Sets to be checked carefully after pre-production.

8) Distribution of jobs, eg., calibration of jigs.

9) Check all engineering change numbers (ECNs) drawings and parts lists.

10) Johnston-Kingston Works informed that colour CRT made by them, which is R.G.B. (red, green, blue) Phosphor, to be changed to High Colour Purity Phosphor.

11) White balance is not changed much and is within range of alignment tolerance.

12) DHHS Parts Control to be implemented as follows:

    a) CRTs and components which are mentioned in Model Change Report Attachment J2 (Component Failure Data Sheet) to be controlled.

    b) Parts which are marked "MD" or "DH" on Parts List (Parts Structure Table) to be controlled.

13) Following equipment are required for Model 2000 rework:

    a) Air-driver - four units (from production dept.)

    b) 22 k resistor - 1,800 units

    c) Sony bond - two tubes (from production dept.)

    d) Frequency counter - two units (one from engineering dept. and one from production dept.)

    e) Gage - six pairs (to be prepared by engineering dept.)

    f) Clips and tapes - A sufficient quantity (from production dept.)

    g) Central processing unit (CPU) - four units (two from production dept. and two from QC dept.)

    h) Hi-pot tester - two units (from QC dept.)

    i) Socket - four units (from engineering dept.)

    j) Packing materials - A sufficient quantity (from production dept.)

    k) AC adaptor (for UK model) - two units (one from production dept. and one from QC dept.)

14) Serial number of reworked models are as follows:-

    2000 Australia   - S. No. 1499 and backwards - 900 sets

    2000 Australia   - S. No. 699 and backwards   - 250 sets

2000 UK	- S. No. 899 and backwards	- 600 sets
	Total	- 1,750 sets

Case Example On Project Management

1) Appointment of project leaders and distribution of work for following models:

   a) Project leader for E 1417 - Mr. P. Performance check by Mr. P and Mr. Q.

   b) Project leader for E 1416 - Mr. R. Performance check by Mr. R and Mr. S.

   c) Project leader for CR 2034 PCM - Mr. K. Performance check by Mr. R and Mr. S.

   d) Project leader for WD 14504 C - Mr. T. Performance check by Mr. T and Mr. U.

   e) End of September - Purchase order for ten kits for WD 1404 C to be opened.

2) Mr. T to look into dual frequency monochrome monitors.

3) First 1,100 pieces of front panel for WD 1404 C from Country X.

4) How are parts for BCD Co. priced?

5) Moulding dies for panels of DP 1942 R and DP 1944 R arrived this morning from Country X.

6) Does CRT from DT 1305 B fit into front panel of DQ 1434 C?

Case Example On Management Of Technical Problems

Staff meeting on following matters:

1) Facilities/Equipment - Sourcing of and training on how to use them. These are required for the new colour monitor line.

2) Drawing System, e.g.:

   i) Revision/Follow-up (ECN) - Alpha-Beta system or our own system?

   ii) Issuing of drawings

   iii) Alignment/checking manual and pre-production report

3) Technical Support, e.g.:

   i) When introducing colour monitor into production line we can get engineer from Country X to oversee.

   ii) We can get engineers and technicians to study the colour monitor.

4) Parts localisation. Sample parts evaluation.

5) Consider CPUs, gages, line flow charts, QC flow charts, jigs, model numbers, etc..

6) Regulations, e.g.:

   i) Safety

   ii) EMI

   iii) X-ray

Case Example On Implementing Of New Project

1) CGA Type - peak-reducing.

2) EGA will become popular - Types of computer monitor.

3) PGA Type.

4) Set up colour monitor production line. Modify into power socket attached convertor.

   To set up colour monitor production line the following equipment are required:

Equipment	Cost
a) One unit of Electex ECG-801 Universal Character Generator	$24,540
b) Three units of Electex 8 Main Power Supply (Body)	$7,410

c)	15 units of Electex TTL PDA Unit (Distributor)	$6,150
d)	Ten units of Switch-box (UK-made)	$5,000
e)	Two units of Minolta TV-2130 TV Colour Analyzer	$18,500
f)	Three units of Kikusui PAD 55 - 10L DC Power Supply	$6,960
g)	One unit of Kikusui 1502 Digital Multimeter	$530
h)	Five units of Hozan HC-21 Degaussing Coil	$400
i)	100 units of BNC Connector P-5	$450
j)	200 metres of Coaxial Cable 5C-2V	$250
k)	Two units of Chassis Performance Check Jig	$8,000
l)	One unit of Equipment Mounting Rack	$2,000
m)	One unit of Tektronic P 6021 Current Probe and 134 AMP	$4,860
n)	One unit of Chain Driven Power Socket Convertor	$18,000
o)	One Installation of Dark Room Form ITC Alignment	$1,500
p)	Two units of Microscope (60 x)	$350
q)	One unit of RCA Pix-350 G Convergence Meter	$3,380
r)	One unit of AC Stabiliser (10 KVA)	$7,100
s)	One unit of Kikusui 60 Mhz Oscilloscope CRT Readout	$5,000
		$120,380

5) 125 units of AC power cord (CP 243C045AII) to be air-freighted.

6) Progress of dual frequency monitors.

7) TW 1270 Z - Packing bag? Stopper? Pads & chassis holders? Palleting? Labels? Name plate artwork to be despatched by 12th. November by DHL.

8) CRT from 1304B fitted with Front Panel from PS 1435 C.

9) Two gages for Dallas Branch Office.

Case Example On Value Engineering And Technical Problems

1) Colour Monitor - Drawings of localised mechanical parts to be ready by the beginning of December. Carry out cost reduction study. Parts localisation. Consider Meica CRTs. Is cost reduction Purchasing's or Engineering's responsibility? Printing on carton box:

   "140 degrees F/60 degrees C" instead of "150 degrees F/66 degrees C". Name Plate: UL/CSA approved material and supplier. Artwork for rating label to be from PQR Co. as soon as possible.

2) DP 2047 S (T) - Beginning of December: Confirm local sample. Drawings of localised mechanical parts ready.

3) Engineering staff to check performance of Meica CRTs.

4) Parts for TW 1270Z:
   (a) Cushions (pads)
   (b) Sliders
   (c) Brackets
   (d) Name Plates - Colour?
   (e) Alpha-Beta Parts
   (f) AC Power Cord
   (g) Signal Cable

(h) Packing Bag

(i) Rating Label - Arrived?

5) Spacers (from Alpha-Beta) to overcome gap between rear cover and bezel.

6) TU 1289 X - Add four sets. These four sets to be produced in Engineering Workshop.

7) 100 sample units to be sent to PQR Co. for performance test next week.

8) Dallas Branch Office brought up following problems regarding TW 1270 Z:

   a) Pedestal does not sit flat on a flat surface.

   b) Centre of bottom side of pedestal touch resting surface and will not allow all four corners to sit flat on resting surface.

   c) When monitor is turned from side to side, pedestal also turned.

       SOLUTION:- Add cushion to centre of base.

9) PQR Co.'s visit to factory as follows:-

   a) 20th. December - Depart DFW for Country B via Thai Airlines

   b) 21st. December - Arrive Country B at 3.30 pm. Stay in New Bliss Hotel

   c) 23rd. December - Depart Country B via Pan Am 101 at 9.00 am for Alpha-Beta. Stay in Hotel Holiday Inn.

   d) 24th. December - Rest

   e) 26th. & 27th. December - Train factory and internal personnel. SQA - manufacture first lot.

   f) 28th. December - Review findings and action plans if needed. Jeffrey Bridges will meet during above schedule with factory people to discuss purchase orders and schedules.

   g) 1st. January to 3rd. January `87 - Singapore. Same basic schedule as in Alpha-Beta with SQA and purchase scheduling.

   h) 4th. and 5th. January - Tandon PDP plant

   i) 6th. January - Departure for U.S.A.

10) Dallas Branch Office representative will also be coming to visit factories.

11) PQR Co. has not announced any delay in schedule and still wants shipments by the beginning of January `87.

12) PQR Co., must first give 1st. Article approval for shipment of first 2,000 units.

13) The five units of colour and monochrome monitor will be used as 1st. Article.

14) The following items for TW 1270 Z are coming in by Adrian Maersk, ETA 2nd. December `86:

   a) 1,100 units of Pad CP 7710004A2l

   b) 1,100 units AC Power Cord CP 713C011Al2

   c) 1,100 units of Knob CP 713C002A21

   d) 1,100 units of Knob CP 700A019Al2

   e) 1,100 units of Bottom Cover CP 701A021A14

   f) 1,100 units of Base CP 700A022A15

   g) 1,100 units of Buffle CP 7000016A10

   h) 1,100 units of Holder Cord CP 761C011Al2

   i) 1,100 units of Bezel CP 701A036A11

   j) 1,100 units of Gate Cap CP 700C105A10

   k) 1,100 units of Clamper Cable CP 595D001A11

   l) 1,100 units of Transistor 2SC2578 - CP 261P034A11

   m) 1,100 units of Transistor 2SC3446 - CP 261P035Al2

   n) 1,500 units of Transistor Fly-back Transformer CP 334P011Al2

15) Complaints from Pulsar Electronics Pte. Ltd. as follows:

   a) Component L555 bearing Code No. 419P006A51 has high rejection rate (two units of rejected CRT socket have been sent here for verification and testing) - 8% to 20%. It burns easily.

   b) Various defects in picture tubes:-

      i) Stray emission effect (observed during horizontal line position) two types - background radiation of either red, green or blue colour during adjustment of white balance.

      ii) Colour arching - about four to five arching lines appearing on picture tube (same PWB set worked OK with other picture tubes), as soon as colour pot is in maximum position.

      iii) Phospherous mask defect - 2% to 3% of tubes have phospherous dust missing from shadow mask. In some, there are some particles blocking holes of mask.

      iv) Faulty tubes - these tubes are faulty internally as there is no snow on screen and only green, or red, with retrace line appearing on screen. In some, arching sound can be heard.

      v) Defocus - 2% of picture tubes have this problem. Pictures are not getting focused properly. In some tubes, focusing is intermitten.

      vi) Rainbow - effect is very much observed in a number of tubes whenever bar pattern is taken at minimum contrast and low brightness. All three colour patterns on right side of CRT give rainbow effect pattern.

      vii) Flashes -

         a) Type I Flash:

           Whenever there are vibrations at PWB mains, there appear horizontal darkbands on the screen. They are very much visible in Low Contrast position.

         b) Type II Flash: Observed at minimum contrast (in Luminance pattern). Horizontal lines appear on the screen. Sometimes, by changing PTS board, flashes disappear. But this does not apply to all cases.(RGB semi-circle radiation on dark position of bar pattern, especially when viewed in dark condition.)

16) Will our other models which have the above-mentioned components also have problems?

Case Example On Technical Problems

1) Alpha-Beta is shipping one monitor and one CRT (polish).

2) Since monitor has non-glare CRT, we are to change the CRT.

3) Dallas is shipping two sets of colour and five sets of monochrome monitor in December.

4) The 1,000 sets of monitor to be mass produced soon are to be with ferrite rings.

5) To obtain 1st. Article of approval.

6) PQR Co.'s 1st. Article of approval.

7) How do we check PQR Co.'s monitors without their CPUs, and with only two units of PC card?

8) To check modified brackets.

9) What happens if after mass production commences on 6th. December, problems are discovered on 100 sample sets?

10) Rework them?

11) What happens if after shipping 1,000 units, problems are discovered on 100 sample sets?

12) Shall we wait for result of sample before shipping?

13) Shall we liase with PQR Co. first?

14) How about getting PQR Co. representatives here to inspect/approve samples of 1,000 units before shipping?

15) Component kits for CTV Model DP 2047 S(T) to be in by beginning of January `87.

16) Performance check/design improvement for WD 1405 F.

17) One set of WD 1405 F with non-glare CRT, ABC Co. logo, and ABC Co. specifications on front panel for ABC Co. in December.

18) Five sets of WD 1405F for PQR Co. in February `89.

19) Name plate, which is based on our factory's specifications, to be changed - no need for "Made in Singapore". One set of name plate to be shown to ABC Co. by end of November.

20) TC 3200 - Three units of sample part to be in by end of November, viz.:
    a)   TF 504 (Fly-back transformer)
    b)   TF 973 (Power transformer)
    c)   Deflection Yoke

21) To get GS Mark = VDE for above sample parts in December or latest January `87.

22) Two units each of TC 3200 final colour samples are ready.

23) Is one set of CPU sufficient for ABC Co.'s TC 3200 mass production?

24) Guest from Sentronics Wire is coming on 13th. November.

25) Lima-Mina Works would produce 300 units of DP 2047 S(T) next month.

26) When are the components for IBM EGA development coming?

27) Have the specifications for UV 1421 B PGA monitor been received yet?

28) PQR Co. wants to postpone delivery of TW 1270 Z, but we cannot allow this because the parts have been ordered and there is going to be storage problem.

29) Going ahead with mass production and hoping that everything goes well.

30) Purchasing equipment for new Australian monochrome monitors and colour monitors.

31) Dual frequency monitor development is already on. New monitor chassis with IBM EGA development is used.

32) Cost reduction study to be carried out will be discussed with PQR people when they are in Country X in end December.

Case Example On Technical Problems

1)   PQR Co. CPUs.

2)   Performance check on PQR Co. models.

3)   To check performance of WD 1404 F. To consider localisation of parts for WD 1404 F.

4)   Setting up of new colour monitor line.

5)   Development of IBM EGA Card monitor.

Case Example On Technical Problem

A)   1942 Front Panel/CP-1344S Door

    1)   Discussion with front panel supplier.

    2)   The following are comments of the front panel supplier:

        a)   Front Panel has been oversprayed. Panel supplier agreed to submit new sample.

        b)   The first lot of 2,500 units has already been specially accepted.

        c)   Colours of Door and Front Panel of Model CP-1344S have a slightly different tone.

        d)   According to supplier, Door and Front Panel have been done in separate processes.

        e)   Supplier agreed to study process.

        f)   In future, supplier would study process, and fix Door and Front Panel together so that difference in tone can be screened out before delivery.

B) <u>TW 1270 Z</u>

1) Customs clearance for five sample sets of TW 1270 Z has been carried out.

2) We will be informed of more test results in due course.

Case Example On Technical Problems

1) CRT from CTV DP 1305 C to be fitted with Front Panel from CTV CE 1543 B to see if the two fit.

2) Strength relief test for AC Power Cord - QC urgently requested. Country X to consider counter-measure. UL regulation for this test cannot be satisfied now.

3) Has one sample set of ST 210 been sent to ABC Co. already?

4) Cheaper CRTs?

5) Pre-production of DP 2047 S(T) in early January `87. To prepare/get ready the following:-

   a) Time Standards Table

   b) Pre-production Reports

   c) Three Component Kits

   d) Instruction Book

6) Johnston-Kingston Works sent over the following:

   a) One case - 54 units of 14" CRT

   b) One case - 72 units of 14" CRT

   With small cushion and with weight reduced by 35% and thickness reduced by 30 mm

7) Check packing case. Check cushion. Check CRT itself. Are all these related to ABC Co.'s problem?

8) Electrical check on CRT:

   a) Capacitance

   b) Purity and convergence

   c) Beam centre

9) ABC Co. complained about scratched CRT. Was CRT passed to ABC Co. at sister company store?

10) CPUs from PQR Co. arriving soon.

11) Have all of TW 1270 Z's problems been solved? How to solve the "top curl" problem?

12) American Branch replied that they hope to clear five sample sets soon.

13) PQR Co. seems to have carried out some more tests.

c) MODULE 3: QUALITY ENGINEERING

Case Example On Incoming And Outgoing Quality Control

1) Dr. Edward's subordinate, Mr. Reginald, his quality assurance manager, would come.

2) Johnston-Kingston Works suggests that we should buy as many parts as possible from Malaysia and Singapore.

3) They want our factory to be the quality assurance/technical centre.

4) Mr. Savill would come in the middle of May. If parts change, design should change.

6) Design of PCB includes the following:
    a) Tuner - DY-FBT } Over 80%
    b) Terminal Antenna
    c) CRT (Cathode Ray Tube)
    d) IC
    e) Optional Components

7) Mr. Lowell decides how they are to help and establishes QA system for local parts.

8) What components can be sent to Country X and other countries?

9) Client A's problem:-
    a) Overshoot Problem
    b) Bandwidth
    c) Hum-bar
    d) Contrast

Case Example On Seeking Quality Approval For New Model

1) 1,000 or 1,100 sets of SB 1000 for VDE section.

2) Matters to look into:
    (a) CRT and major components.
    (b) Housing colour, logo and carton box design - these are under preparation.
    (c) Radio interference - GS mark.

3) TUV (KEC) & VDE = GS mark

4) Purchase order to be received by 4th. September for November production.

5) All correspondences to be directed to new address of Canada Sales Office by 21st. September.

Case Example On Quality Management

1) Logo for SB 3000.

2) Model No..

3) Drawings to be issued end of October. No problem.

4) Instructions Book to be based on our standard request from Alpha-Beta.

5) Size of packing case to be same as that of model.

6) Wordings on packing case.

7) Buy EGA (Enhanced Graphic Adaptor) colour monitor for ANC Co..

8) Power Switch - green colour is to be used (red is not acceptable by VDE and VDA).

9) CRT - Non-glare screen is wanted.

10) Power cables.

11) Will sell monitors in PAL (Phase Alternating Line) areas.

12) Electrically, we can follow completely.

13) By mid-December, parts for pre-production to come in first (two sets). By early November one complete sample to be ready for evaluation and picture "cut - off" test.

14) Apply for VDE approval - 50 sample sets of SG 3000.

15) Submitted sample to TUV. Three stages are involved:

    a) Radio Frequency Interference Test (Mid-October)

    b) Ergonomics:

        i) LED (Light Emitting Diode) must be green (check CRT).

        ii) Actual colour of cabinet must be given. (HIJ Co. to provide two units.)

        iii) On label behind, there must be statement that x-ray leakage is less than 27,000 volts (PTV).

    c) GS-VDE (Indicate fuse position, just after transformer):

        i) Parts samples - three units each of power transformer TF 502, power transformer TF 701, deflection yoke (DY) and flyback transformer.

        ii) Power switch to be (preferably) VDE approved.

        iii) It is not necessary to submit power transformer, TF 971, for VDE approval. Primary/secondary insulation type, which is same as UL type, will do.

16) Application to VDE to be made by November, and latest, by December.

17) All samples of power transformer to be ready by end of October.

18) ABC Co., Dallas, sent PROM and diskette to our American Branch. They are revised because the last one had problem with the software.

19) PROM and diskette have been sent to ABD Co., Singapore, a few days ago.

20) ABC Co., also sent three types of drawing. Are they requested by our mechanical engineer? To check with mechanical engineer.

21) Jim Reefolds from our American Branch to request from Richard Lumky:

    i) Suggestions for software or procedural improvements.

    ii) Any gaging issues.

22) Tandon PCs (Personal Computers) at Alpha-Beta office - two units are working while three units are out of order. In our factory, one unit is out of order.

23) We have to request for new PCs (for testing purposes).

24) Colour monitor is to be ready by 25th. October.

25) Monochrome monitor to be ready one week before ABC Co. people visit our factory on 11th. November.

26) TW 1270 Z to be shipped before December.

27) 98 units of colour monitor sample have been sent from our Alpha-Beta office to our American Branch.

28) They have passed EMI test based on FCC Class B requirement on 25th. September. So, we would get grant before ABC Co. receives the first "mass production" lot.

29) Items that passed EMI test are as follows:

    a) PC Unit - PC-It (Sperry) compatible with IBM PC-AT.

    b) Graph Board (supplied by ABC Co.) - without the shielding cover that we received from our American Branch.

30) The following are counter-measures:-

   i)   Conductive point inside cabinet is for colour monitor and not monochrome monitor.

Colour Monitor	Monochrome Monitor
ii) Shielded cable	Shielded cable
iii) Signal cable - double shielded with two ferrite rings	Signal cable - double shielded with one ferrite ring
iv) Shielded video cable between main PCB and CRT PCB	
v) Ferrite ring at lead of CRT coating earth	

31) When ABC Co. conducts source inspection, 1,000 units may not be completed on time. Discuss with Alpha-Beta office.

32) ABC Co. to train Intek staff in Country X and to train their staff in Singapore.

33) Source inspection to be carried out by ABC Co. representatives later.

34) ANC Co. visiting schedule as follows:

20th. October - Arrival in Alpha-Beta

30th. October - Inspection

31st. October - Inspection

1st. October - Shipment

35) Due to delay of cabinet supply, monitor assembly in Singapore may have to start on 8th. November, and inspection by ABC Co. may have to be carried out on 14th., 15th. and 16th. November.

36) Is mass production to be carried our on 4th., 7th., 8th. and 9th. November? To confirm this point by 18th. October.

37) DP 1942 R

   i)   Cabinet (720B487-2) outer size to be big.

   ii)   PVC sheet - Gloss degree is too much/colour is different.

   iii) Panel front - Consider number of staples and position.

38) 44 CC 22 HC - brand name to be changed to USA version (from Canadian version).

39) ECN (Engineering Change Number) will be issued after confirmation.

40) New transmitter (Remocon Transmitter) will be supplied.

41) The labels, "Ant" and "TV - Ext", are to be excluded.

42) Mass Production of 44 CC 22, HC - 1,500 sets to be produced during the period, 14th. November to 18th. November.

43) Quality Assurance Dept. has asked SISIR (Singapore Institute of Standards and Industrial Research) to advance 1,500 pieces of CSA label on 9th. November.

Case Example On Application For Quality Approval

1) Alpha-Beta Branch checked with Dallas and Michigan branches on the following:

   a)   VDE B - EMI requirement

   b)   GS mark (safety requirement)

   c)   Any approval obtained for ABC Co. PC? (VDE - f mark, TUV - RFI mark, self - control?)

2) If not, ABC Co. to apply for VDE B APPROVAL.

3) Our company to apply for GS mark. (TUV GS mark and not VDE GS mark.)

Case Example On Quality Management

1) <u>XY Packaging And Our Company Good Manufacturing Practice Scheme</u>

    a) XY Packaging will study scheme and let us know outcome of discussion as to whether to accept scheme.

    b) If decision is positive, our Quality Assurance Manager will approach SISIR to provide initial consultancy services.

    c) "Defective" problems, e.g., fishtailing, creasing and overslotting, brought out.

    d) XY packaging representative said that all carton boxes for our company will be stitched rather than glued.

2) Five samples of TW 1270 Z shall be aired on 8th. December `86 or earlier (enough time for EMI testing).

3) 1,000 units of mass produced TW 1270 Z to be shipped out on 22nd. December or earlier.

4) Name Plate to be L-shaped.

5) Modify Polyfoam.

6) Models 1942, 1944, 1954 and 35 CC 22:

    Transistor 2SC812B-F-H, Part No. 261D168A21, at location Q734 on PCB (Printed Circuit Board), is to be replaced by Transistor 2SC2803 F-H, Part No. 26IP348A70.

7) PG 1000 HK and PG 2000 HK to be without Singapore label and with cable.

8) PQR Co. Monitor FCC Issue (TW 1270 Z and QS 1209 D):

    a) FCC acquisition schedule may be "End December" or earlier.

    b) 1,000 units of colour monitor already shipped out and now on the way to U.S.A..

    c) Expect acquisition schedule to be on 20th. December, when goods arrive in the U.S.A..

    d) If approval is not possible by 20th. December, our American Branch is to hold shipment to PQR Co..

    e) FCC regulation does not permit importer's shipment to their customer without their (FCC) approval.

    f) PQR Co. has been promised that goods will be delivered to them by 22nd. December or so, so that they can distribute the goods to the outlets by January `87.

    g) We now seem to be at a deadlock.

9) Questions asked and points raised:

    a) Will PQR Co. start selling monitors from January `87 through their sales channels (retail stores)?

    b) If this is so, we may take the risk of shipping the goods to PQR Co. without FCC acquisition.

    c) If sales start later than that, FCC acquisition to give interim report on FCC around 10th. December. We can discuss issue at that time.

Case Example On Quality Management

1) <u>TW 1270 Z Mass Production Interim Report From QC</u>

    a) Front panel filed/cut at both top corners.

    b) Front panel has three pieces of plastic material fixed to its top interior to minimise <u>height</u> differences that occur between front panel and back-cover.

    c) Back-cover has two gum washers fixed to its bottom screw points to minimise gap differences between front panel and back-cover.

    d) PWB-main to be supported by PWB-holder and cushion pads to prevent PWB-main and back-cover cracking, upon drop test.

e) Gap of top cushions to be filled up, to prevent back-cover from cracking upon drop test.

f) Packing of monitor units (packing sheet) - the current method of packing requires much time.

g) Strength relief test - Failed. Urgently request Country X to consider counter-measures. UL regulation for this test presently cannot be satisfied.

h) High voltage regulation - At 98 V AC, raster is slightly shaky and small humps/waves appear. Engineering Dept. currently studying matter (Country X to look into matter.)

i) Brightness drop (about 10 to 20 Nits) - Engineering Dept. has already instructed production to align beyond 150 Nits to off-set the difference.

j) Raster shift - Deadlock situation. Gages off-set has been changed a few times. Raster shift is most.

2) Models P32 and P37 for UK market. 5" and 9" - 1,000 to 2,000 units a year. 5" model to have polish type/glare type CRT. 9" model to have non-glare type CRT.

3) Nixdorf sent one sample of scratched CRT here.

4) TC 3000 - Bezel, back-cover and cabinet to be of same colour.

5) Can provide rating label in German.

6) Nixdorf logo to be silkscreened or pasted? Red colour for lights will be given.

7) Packaging details.

8) Printing film.

9) Label and logo - Will send in January `87.

10) Instruction Book - Our factory can prepare English version.

11) Kind of material for rating label - paper or cloth? GS mark (provided or print ourselves?).

12) Samples will be ready by end of December.

13) Warning label for X-ray and servicing - to be in English and French according to UL and CSA regulations.

14) Parts for testing have already been sent.

15) Warning Label change for TW 1270 Z - Label to include "And The Same Rating Fuse".

d) MODULE 4: SOURCING OF PARTS

Case Example On Sourcing Of Parts

1) Johnston-Kingston Works also comprises of the Lima-Mina Factory.

2) Springleaf Works manufactures refrigerators and air conditioners.

3) Mr. Warner Lester works in the Continental Semiconductor Works.

4) The factory houses the Central Research Laboratory and the Product Development Centre.

5) Alpha-Beta Works now manufactures monitors.

6) Johnston-Kingston Works now makes mainframes that rival IBM and Univac.

7) To send purchase order to the Production & Material Control Section of the Overseas Licensing Service (OLS).

8) The HQ of Continental Works is at Burnstedt Computer Centre at Country Y.

9) To send purchase orders to various suppliers at Country Y.

10) The lead time is four months. Shipment takes one to two months.

Case Example On Purchasing Of Parts

1) Localisation of parts, e.g., cathode ray tubes, tuners, fly-back transformers, remote controls, lead wires, deflection yokes, etc..

2) 85% of parts can be localised.

3) Integrated circuits (ICs) and other key parts to be from Country Y and Country Z.

4) Resistor R419 of Model 1903 is giving problems.

5) Torgue meter to check torgue at cathode ray tube.

6) Box spanner for tightening screws on sets.

e) MODULE 5: SALES ENGINEERING

Case Example On Sales Engineering

1) Following facsimile to be sent out to Zenith Company:-

   "This fax serves to inform you that your proposal for the location of the data code on the monitor PWBA has been accepted.

   As per your proposal, it shall be a 6-digit code in the month-day-year sequence."

2) i) Raster height for monitor to be 163 mm ± 4 mm.

   ii) Raster width to be 216 mm ± 4 mm.

3) Preparation of jigs before production starts.

4) Humidity test.

5) To recruit work study officer.

6) QC to check Model 1954.

Case Example On Sales Engineering

EFG Co. 12" Black/White Monitor

1) EFG Co. agreed that video input and output with 75 switch is not absolutely necessary, i.e., we can supply them monitors with attached signal cables.

2) Can most probably use Sperry-type cabinet without any change.

3) The following questions are to be answered:-

   a) Can we use Sperry cabinet or is there any copyright?

   b) Can we modify PCB (Printed Circuit Board) to analogue type?

   c) Can time frame to make sample based on the above base be indicated?

   d) Can quotation based on the above and the requested quantity from EPG Co. be given?

4) The above information are required by 8th. October.

5) Facsimile from Mr. Remick on exchange rate of Singapore dollar which is as follows:-
   S$2.225 = U.S.$1.00

6) Mr. Tim Rogers, Manager, Tube Engineering, California Branch, to give approval of CRTs by early November.

7) To check luminance of three portions of CRT:

   a) Top Portion - 86 Nits

   b) Centre Portion - 43 Nits

   c) Bottom Portion - 22 Nits

   Luminance has to comply with FCC (Federal Communications Commission, U.S.A.) requirements.

8) Drop test on 20 sets. After drop test, two sets were found to have cracks. Check on contrast/brightness showed no problem. PCB in one set broke.

9) To carry out drop test first in future.

10) SW 1298 - Local mold for this model arriving on 4th. November.

11) DP 1942 R - Front panels and back-covers coming from California Branch. First 1,500 sets also from California Branch. Molds also coming from Branch.

12) SU 1260 Z - Front panels and back-covers for model from Country X.

13) Hotel reservation at Holiday Inn, Country X, for Mr. Tim Fast from Dallas Branch. Period of stay is from 19th. November to 24th. November. Checking out on 25th. November.

14) Mr. Roy Hunt, inspector of ABC Co., to meet Mr. Tim Fast at Alpha-Beta.

15) 19th. to 24th. November - Alpha-Beta Factory inspection by Mr. Roy Hunt on first production lot.

16) Messrs. Roy Hunt and Tim Fast will be at our factory from 27th. November to 29th. November for discussion. There will not be factory inspection because model is not produced at that time.

17) Johnston-Kingston will ship out following new CRTs:-

    a)   380LFC22ETCO2A - Purchase Order. 1345S-24 - ETD 21st. November `86 - 9,000 units (7,080 units - new type. 1,920 units - present type.)

    b)   380LFC22PTCO2A - Purchase Order. 1417 (Q) A - ETD 21st. November `86 - 2,000 units (1,736 units - new type. 264 units - present type.)

18) Both new types can be used as usual. Identification of new type: "Z" after usual type number on CRT label and on carton box.

19) R.G.B. Phosphor to be changed to High Colour Purity Phosphor.

20) TW 1270 Z monitors to have dual frequency.

22) Mr. T. Joseph, inspector of ABC Co., would be in factory on 28th. and 29th. November. He is to train members of Intertek and Q.A. staff.

    He requires:

    a)   Software

    b)   Overlays

    c)   A light meter

23) If purchase order is issued by end of November, 400 units of TW 1270 Z can be shipped out on December `86 and 400 units on January `87.

24) Labels, packing cases, etc., to be different from earlier lot.

25) The following are results of drop test on TW 1270 Z:-

    a)   Panel cracked (at top two corners)

    b)   PCB cracked - counter-measure needed (ECN?)

    c)   Contrast is out

    d)   Carton box quality problem

    e)   Label is missing

    f)   Power cord is loose

    g)   Adaptor is damaged

26) The following are problems of TW 1270 Z, which are to be solved:-

Mechanical

    a)   Gap between front and back covers

    b)   Name plate is torn

    c)   MC connector is too high

    d)   Back-cover screws not tightened

    e)   ABC Co. name - double printing (on bezel)

    f)   Rear cover lid cracked

    g)   Mesh sheet loose (at back-cover - feed back to Production and QC)

    h)   Metal frame problem:-

        i)   Gap between CRT and bezel

        ii)   Gap between bezel and back-cover

Electrical

a) Focus problem

b) There is top curl

c) Bright spot

d) Unstable picture (vertical jitter)

e) B4 range to be less than 13 volts (13 volts to 14 volts for x-ray)

f) Check protector circuit

g) Short regulator

h) Check power transformer

i) Pre-heat sets and then let them age for 10 minutes

j) Further check on a few other sets

27) Quality of rear covers: Plastic parts from Alpha-Beta. 1,500 sets - painting texture. 1,000 sets - molded texture.

28) There are black spots/burrs - a lot of stains.

29) When are CPUs arriving from ABC Co.?

30) How to overcome above problems? Modify CRT holder - holes to be smaller for proper mounting. Modify mounting jig.

31) Permanent counter-measure for PWB - main mounting is required.

32) Consider new molding die.

33) Consider using sliders and cushions.

34) Packing:-

a) Packing sheet is too complicated.

b) Stapling at bottom of packing case.

c) Drop test was carried out with just tape at top and bottom. Based on test result, it is found necessary to have stapling (with 24 staples) together with taping.

d) Consider using packing bag.

35) Gages for display adjustment - new gages for 1,000 sets of TW 1270 Z.

36) Signal cable is too small for damper. Speak to supplier.

37) For safety regulations and quality assurance purposes, drop test, as well as life test, will be conducted on one set of each production lot for all UL and CSA models with effect from November `86.

Case Example On Sales Engineering

12" black/white monitor for LMN Co.

1) Our sales office here has no exclusivity of cabinet covers for European market with Sperry.

2) We can use Sperry-type cabinet, but not its logo.

Case Example On Sales Engineering

1) Status of PQR Co. first article of approval (for 98 units).

2) Change of testing procedure.

3) Difference of signal timing.

4) "B" shadow issue.

5) Dallas Branch Office to reply on above.

6) <u>TW 1270 Z</u>

i) Name Plate - Material to be mirror coat 85 gm. It must be:-

   a) UL/CSA approved (recognised) material

   b) Made by recognised manufacturer

ii) Bar Code Label - Alpha-Beta sent reader program for Code 39. Alpha-Beta checked program and found it to be alright. Engineering Dept. to confirm program is fine.

iii) Pad CP771D004A20 will be sent with cabinet.

iv) Plastic Covers - Alpha-Beta approved 60 degrees celsius, 90% RH (Relative Humidity).

v) Humidity - PQR Co. has no comment on humidity probably because products may go quickly to customers and will not stay in storage. PQR Co. has approved humidity of 95% concerning the 2,000 units.

vi) Recovery Of Sag - There is none. Once top panel sags, it will not go back to normal position.

vii) Return Of Monitors - PQR Co.'s request to return monitors is reasonable. They can return monitors that have any sag problem to our American Branch. Our mechanical engineer to note this point. When PQR Co. Inspector is in Country X, he will discuss sag problem, especially when sagging distance is more than 1 mm.

viii) Head Office in Country X will determine how to service PQR Co. monitors by discussing this with our American Branch sooner or later. The matter of the replacement of sagged cabinets will be included in the discussion. American Branch will be asked to replace cabinets at their facility.

xi) Printing On Carton Box - Carton box is to show "140 degrees F/60 degrees C" instead of "150 degrees F/66 degrees C".

x) Base - To be blue in colour and same as background of carton box. To prepare additional carton box labels (at least 1,050 pieces). All carton boxes to be labeled.

xi) Improvement Of Plastic Material - Alpha-Beta investigating this issue, and will advise as soon as they have something. It is almost impossible to fully meet PQR Co. specifications. May be asking for some meeting half-way.

xii) PC Repair - In order to put all devices in order, the following are needed:-

   a) Graphic Board (Three units. Other EGA type?)

   b) Power Supply (One unit)

xiii) Ferrite Ring - To clear FCC regulated level, ferrite ring is not enough by itself. Some other counter-measures like electric conductive painting are necessary.

xix) PQR Co. wants to look at two units of final sample. Two units of final sample (only colour monitor) to be shipped out of factory by 24th. October, ETA U.S.A. 27th./28th. October. The samples will be somewhat different from the mass-produced sets in that the FCC ID will be erased because even by that time FCC approval will not have been obtained.

xx) Overlays (Templates) - Need to provide PQR Co.. Will hand over to PQR Co. Inspector a few pieces of overlay. Our American Branch can get one of these.

xxi) Distortion Inspection Procedure (Colour) - Will comment later. Engineers would prefer cross-hatch. Can discuss with PQR Co. Inspector when he is in Country X. What does "2 ft." mean? Some data will be sent over tomorrow which may clarify everything.

xxii) Alpha-Beta discovered problem which caused molding parts (especially bezel) to change in shape at 66 degrees C, 95% R.H.. So they proposed to PQR Co. for 1st. lot (1,000 sets) to change the label indication on the packing case shown below:-

---

140 degrees F/60 degrees C

Label CP851D270A14

xxiii) EMI, VDE B - General. PQR Co. to apply for VDE B as a system (personal computer and monitor).

xiv) Timing chart of TW 1270 Z - Can go with timing chart shown in specification.

xv) Our company to apply for TUF (RFI mark) as standalone (monitor only). Application fee is US$150,000.

   a)   Can obtain TUF in Country X, and this is quickest way.

   b)   GS mark (safety regulation).

   c)   TUV - GS mark.

   d)   PQR Co. visit schedule to Country X.

   e)   Approval on cabinet colour sample.

   f)   Artwork for rating label. To ask Dallas Branch Office to send us as soon as possible.

   g)   RX version mass production schedule.

7)   Jigs for 1942?

8)   Progress of dual frequency monitors?

9)   Colour monitors - Component kits? Equipment?

10) TW 1270 Z - Packing bag size? Name plate colour? Jigs?

11) Quality Control Circles (QCC)

12) Colour monitors for PQR Co.

Case Example On Sales Engineering

1)   PQR Co.'s visit to Country X is not yet confirmed.

2)   Cabinet colour samples - When Mr. Eric Peterson, PQR Co. representative, was in Osaka for tooling meeting, he said he had not received colour samples for authorisation. Both our company and PQR Co. have forgotten about it.

3)   In a week, colour samples were sent to PQR Co.'s New York Branch as per Mr. Peterson's request.

4)   PQR Co. replied that colour matches are slightly off.

5)   Alpha-Beta Marketing Dept. informed Mr. Peterson that we cannot help but go with the present colour as far as 2,000 units are concerned, but received no answer on this point.

6)   Alpha-Beta Marketing Dept. will send new colour samples in a few days' time. They consider that PQR Co. actually accepted current colour for the 2,000 units (no reply from PQR Co. about these 2,000 units).

Case Example On Sales Engineering

1)   Pedestal Issue - 98 units of TW 1270 Z will be evaluated as part of 1st. Article of Approval except for cabinet, because cabinet tooling is still tentative at this point.

2)   The cabinet was to be evaluated with five units, which are equipped with final cabinet (final tooling). We believe that PQR Co. people are aware of this point (end brackets).

3) PQR Co. representatives to visit Alpha-Beta Works. Holiday Inn Hotel reserved for three persons.

4) Shipment of 1st. 1,000 colour monitors:-

   a) Shipping 1,000 units out of Works on 2nd. December.

   b) Shipment to be FOB (Free On Board) Country X. Date of leaving (ETD) will be confirmed soon. It is tentatively 7th. December.

5) Evaluation result of 98 units:-

   a) PQR Co. evaluation is delayed, it seems. Their comments may not reach Country X before shipment.

   b) PQR Co. has to take up 1,000 units anyway.

   c) Problems, particularly technical problems, will be discussed with PQR Co. people when they are in Country X in End December.

   d) Hopefully, PQR Co.'s comments can be applied from second lot.

6) TZ-0911 PU - To add seven sets on February `87.

7) 2,000 units of DT 1315 KF (based on DT 1314 KF) to be for April `87 mass production.

8) Critical Component Lists and Material Identification Lists of new U.S.A. models, PG 2000 and TG 2500.

9) The setting criteria for Dielectric Strength Test is also the same as for other U.S.A. models, i.e., 1.25 kV with trip current set at 3mA (For Materials Dept's and Production Dept's action).

Case Example On Sales Engineering

1) PQR Co. Problems - Distortion: geometry and contrast. AC power plug.

2) PQR Co. people coming in early January `87. To be confirmed.

3) Scratches on CRT for 2038 (ABC Co.).

4) Cost reduction study for colour monitors. Value analysis.

5) Progress of development of IBM EGA Card monitor.

6) PQR Co. CPUs coming in End December.

7) Component kits for 2034. Pre-production to commence in Beginning January `87.

8) Check PQR Co. monitor again. Take down data with one set from production line. Check swivel mechanism.

9) Any reply from American Branch? Are there any more results to be relayed?

10) Factory equipment calibration to be carried out.

11) Cushions are to be used for noise reduction in new production line.

12) DT 1315 KF (based on DT 1314 KF) for April `87 mass production. Mechanical aspects are OK.

13) How about PQR Co. signal cables? How about distortion and contrast? Consider the following:-

   a) Focusing

   b) Convergence

   c) Brightness range

   d) Display area

   e) Display centering

   f) Contrast

14) Two pieces of template to be given to PQR Co..

15) 20th. December - One sample of ST 210 (with cable) to be sent to ABC Co..

16) Localisation of parts for models F 3014 T and F 3015 T.

17) Checking of Meica CRTs is in progress.

18) CPU for ABC Co.'s PH 3100 - Will one set be supplied to us? When is mass production for this model?

19) Is the first PQR Co. performance check completed already? Are there any more checks being carried out? Is there any magnetic field interference, and, hence, distortion?

20) Magnetic field interference is apparent during initial "switching on" period of power supply. Production to pay special attention to setting their magnetic chamber field during such interval.

21) Miscellaneous - Request PQR Co. to provide enough CPUs for production use and Q.C. inspection. Currently, "National" video generator and our box generator are being used. Quality level obtained via each type of equipment is inevitably different.

22) PQR Co. Software:-

    a)   Line width specification is not clear.

    b)   Raster Corner Angle is not found in the initial contract specifications.

    c)   Whether profile and parallelism are similar or not?

    d)   Raster centering: 86 Nits or minimum raster size?

    e)   Reverse/maximum brightness cannot be obtained with PQR Co. software.

f) MODULE 6: TECHNICAL COMMUNICATION

Case Example On Technical Communication

1) Molding die for TW 1295 to be made as soon as possible.

2) Mass production of monitors, Model TS 2000, TU 2500, TW 1270 Z.

3) First lot, 1,100 pieces, of front panel for TW 1270 Z from Country X.

4) Chasis, preamplifiers and key-boards for DP 1942 R and DP 1944 R to be from sister company.

5) DP 1942 R and DP 1944 R to be mass produced on 18th. October.

6) Mr Thomas from Lima-Mina Works here on 17th. and 18th. October.

7) Mr. Chamberlane here first week of October to discuss colour monitors.

8) Mr. Mitchell here from 2nd. to 5th. October to discuss Johnston-Kingston related problems.

9) 250 sets of DP 1944 R to undergo aging and 100 sets to be put aside as sample, giving a total of 350 sets.

10) According to Mr. Robert Leach, Australian Branch, "Made in Singapore" label is not necessary for Australian market.

11) TS 2000's and TU 2500's CRTs to be same colour as ABC Co.'s.

12) Model CP 2103 E for Indian market.

13) DP 1942 R to have panel control (part no. 702B475).

14) Mr. Parker to inform mechanical engineer about new back-cover.

15) Two pieces of battery, instead of one, for Models 35CC12, 1942, 1944, 1957 and 1959. Refer to ECN (Engineering Change No.) BY-1403 from Johnston-Kingston Works.

16) 100 units of TW 1270 Z monochrome monitor to leave Singapore on 7th. October, FOB (Free On Board).

17) Additional 20 sets of TW 1270 Z to be obtained from first lot of 1,000 sets to be mass produced in October.

18) Discussion on following work problems:-

    a) Work-load

    b) Communication with other departments/Johnston-Kingston Works

    c) Equipment and tooling

    d) Cleanliness and orderliness - premises to be cleaned at all times as there are many visitors

    e) Circulation of information internally and to other departments - information often reaches person concerned too late

    f) Recruitment of new pre-production operator

    g) Production operations - work study to be carried out to improve factory layout and operations

    h) ECN/Drawings/Part Lists. Can system be improved?

    i) Introduction of new models

    j) Welfare - picnics, games, etc., once every two or three months

    k) Recruitment of process technician

Case Example On Technical Communication

1) Chicago Sales Office replied to our QC Dept. as follows:-

    a) It is alright for us to release shipment. They have received UL (Underwriters' Laboratories) acknowledgement, which is enclosed.

    b) UL Engineering Department has completed all their work. Revisions are being finalised.

c) They (Chicago Sales Office) are preparing Notice of Completion for project and will fax copy and mail original of Notice to us.

2) <u>PQR Co.</u>

   a) FTZ (RFI) will be ready by mid-October

   b) Ergonomics:-

      i) LED must be green (Final Nixdorf colour samples - two units each)

      ii) X-ray > 20 kv for PTV

   c) GS/VDE:-

      i) Fuse position is unnecessary:

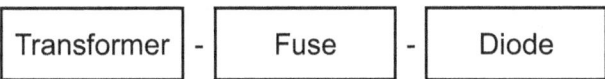

      ii) Sample parts to be sent to GS/VDE by early November as follows:-

Three units each | TF 503 (Fly-back Transformer)
TF 972 (Power Transformer)
Deflection Yoke

      iii) Power switch is not necessary

      iv) The primary and secondary windings of the Power Transformer are to be "isolated" type

      v) "GS" mark would be obtained by November or, latest, December

3) For Models DP-2345 S, 45CD23, DP-1445 S, 1848, 1849 and 1850, ICs (integrated circuits) are to be changed, from IC M51383P to IC M51411 SP.

4) AGC (Automatic gain control) characteristic is to be improved, as per ECN 88/006/ECN 094.

5) <u>Model DU - 1418 PB</u>

<u>Probe on two sets before</u>   <u>Probe on 2 sets after</u>

<u>beam current is adjusted</u>   <u>beam current is adjusted</u>

(RF-AGC is to be adjusted to 0.8 volt, peak to peak)

i) Serial No. 2114: 1.7 volts peak to peak (before beam current adjustment)	i) Serial No. 2113: 1.5 volts peak to peak (after beam current adjustment)
ii) Serial No. 2115: 1.5 volts peak to peak (before beam current adjustment)	ii) Serial No. 2116: 1.4 volts peak to peak (after beam current adjustment)

Result of Probe: Beam current before adjustment is alright.

6) <u>UL For ABC Co. TW 1270 Z</u>

UL to complete investigation by 21st. October. Mr. Kingsley will contact UL if they can pick up temporary procedures from UL and send them to a Singapore inspection centre (SGS?) directly by courier.

7) Mr. Leslie Richard (Compliance Engineering/Technical Liaison) replied Mr. Simpson (Export Manager - Display Monitors) that UL will issue temporary procedures and send them to Country X by courier (cannot fax). IPIs may not be required. He will check with UL and advise about IPIs later.

8) TS tables, ALI assembly and parts drawings, family trees, pre-production reports, sample sets, five component kits and product manuals for colour monitor, Model WD 1405 F, have been sent to us.

9) End of October - One sample set of WD 1405 F is to be ready.

10) November - One sample set of WD 1405 F with non-glare CRT.

11) January 1987 - Pre-production of five sets of WD 1405 F.

12) March 1987 - Mass production of WD 1405 F.

13) ABC Co. - UL to complete investigation by 21st October. (Mr. Leslie Richard will check with UL). IPIs may not be required.

14) Some technical points to note:
    i)   Take out unnecessary things
    ii) Check adjusting jigs (Where to store them?)

15) Introduction of Model 1288 Z - molding die is coming in on 16th. October.

16) Engineering Dept. to pass customer specifications to Q.C. and Production Departments.

17) Cabinets for TW 1270 Z are coming in by air in November. Other materials to come in by air too.

18) 5th. November - Mass production of TW 1270 Z.

19) 17th. and 18th. October - Mr. Charles is coming from Alpha-Beta.

20) Schedules are as follows:-
    i)   1288 Z - Introduction on 23rd. October
    ii) 1942 - Drop test on 25th. October
    iii) 1944 - Drop test on 1st. November
    iv) TW 1270 Z - Mass production on 5th. November (to be confirmed by 18th. October)

21) CRT of 1304 B is to go with panel of PS 1435 C.

22) Gap between CRT and bezel - Add new hole; we are to check before modification.

23) To prepare packing sheet/packing bag.

g) MODULE 7: MAINTENANCE AND CALIBRATION OF FACTORY EQUIPMENT

Case Eample On Factory Equipment Maintenance And Calibration

1) Maintenance of factory equipment should be once a year.

2) Calibration of factory equipment to be once every six months.

3) Room temperature (air condition) should be around 23 degrees C.

4) Humidity should be below 70%, and measured with RH meter.

5) Calibration charts should be put up on wall to indicate status of calibration of equipment.

6) A schedule of calibration should be included.

7) There should be a filing system for calibration reports.

8) The accuracy for factory equipment should be: 1 is to 4.

9) Training should be provided for technicians on maintenance and calibration of factory equipment.

10) There should be liaison with Quality Assurance Dept. and SISIR (regarding calibration).

11) The following are equipment which are used for calibration:-

1st. Standard

(a) DC Meter Calibrator YEW 2560

(b) AC Meter Calibrator YEW 2558

(c) AC High Voltage Meter Kikusui 109 - 10A

(d) Standard Resistor GRP - 1434

(e) H-Frequency Voltmeter ML 69A

(f) Q Meter BP - 4342 A

(g) Oscilloscope Calibrator Leader LOC 7000

(h) Frequency Standard HP - 86568

(i) AM/FM Modulation Degree Meter - Marconi TF 2300

(j) Distortion Meter - HP - 334 A

(k) Standard Magnet

2nd. Standard

(a) Digital Multimeter - Kikusui 1400

(b) Bridge YEW 2755

(c) Frequency Counter TR - 5823

(d) Luminance Meter

(e) Hi-Pot Calibrator T05 -1200

(f) Gauss Meter

Addition To AC High Voltage Meter

HV Generator Kikusui 8652

Case Example On Factory Equipment Maintenance And Calibration

1) Maintenance of factory equipment - Once a year should be sufficient.

2) Calibration of factory equipment - This depends on the degree of accuracy needed. Calibration can be carried out once every six months or once every year.

3) Room temperature - Air-condition is important. Temperature should be around 23 degrees C.

4) Humidity - Should be below 70%. Preferably equip with RH meter.

5) Voltage problem (Sometimes voltage is too high and accuracy of test equipment may be affected) - Instal line adaptor.

6) Interfacing of test equipment with computer (computer-controlled).

7) Certain sensitive equipment are to be free from interference.

8) Certain equipment such as colour analyzer can be calibrated by the Building Science Department of the National University of Singapore. (Person to contact: Mr John F. Pickup)

9) It is possible to schedule calibration by computer.

10) Filing system, with index for calibration reports.

11) Accuracy required - 1 is to 4. (SISIR practises 1 is to 10.)

12) Calibration can be carried out at SISIR (Singapore Institute Of Standards And Industrial Research) or SEEL (Singapore Electronic And Engineering (Pte.) Ltd.).

13) Equipment to be calibrated at SISIR are as follows:-

    a) Electrostatic Voltmeter

    b) Distortion Analyzer

    c) Spectrum Analyzer

    d) LCR Meter

    e) Q Meter (one unit in plant only)

    f) X-Ray

    g) Colour Analyzer

    h) Thermometer (one unit in plant only)

15) The following is a list of calibrating equipment:-

Calibrating Equipment	Equipment To Calibrate
DC Meter Calibrator	
Digital Multimeter	
AC Meter Calibrator	Ammeter & Voltmeter
High Voltage Generator	
AC High Voltage Meter	
High Frequency Voltmeter	
Frequency Counter	
Frequency Standard	Signal Generator
AM/FM Modulation Degree Meter	
Distortion Meter	
High Frequency Voltmeter	
Frequency Counter	
Frequency Standard	Oscilloscope Generator
Distortion Meter	
High Frequency Voltmeter	Electric Field Strength Meter
Oscilloscope Calibrator	Oscilloscope

Standard Resistor YEW 2755 Bridge	Insulation Tester
AC High Voltage Meter High Voltage Generator Hi-Pot Calibrator	Hi-Pot Tester
Frequency Counter Frequency Standard	Frequency Counter
Luminance Meter (Calibrated in HQ)	Luminance Meter
Gauss Meter Standard Magnet	Gauss Meter

16) To order one set of tools and five tool-boxes for Engineering staff.
17) New CRTs should be checked as follows:

Mechanical

a) Fitting to the cabinet
b) Deflection Yoke putting
c) CP - Assembly putting
d) CRT Socket matching

Electrical

a) CRT capacitance
b) Adjustability (Purity/Convergence)
c) Beam Centre (1.5 mm?)

Today, the engineer armed with an engineering degree is just another student in the University of Life-Long Learning. He cannot afford to be complacent and rest on his laurels. The world is undergoing great technological advances and the scientist-engineer must be keen on keeping abreast of all the changes in technology.

The scientist-engineer has to really have a great interest in his field of specialisation if he wants to find out much about it and advances in it. Without this special, deep interest in the science, he could not expect to be able to keep up with the field. This is the most critical aspect of the scientist-engineer's make-up.

Secondly, he has to have access to scientist-engineers of similar interests so that they could test each other's ideas via discussions, seminars and research. They could even collaborate in research work, which would make the search for new knowledge a much more interesting one. "Two Heads Are Better Than One" is indeed here a truism.

Thirdly, these "hi-tech" personalities should be prepared to publish their findings or researches frequently and freely. The exchange of ideas through technical or scientific journals is an indispensable source of scientific inspiration as well as information. Today, such journals abound and are to be found in public libraries and libraries of academia.

Even scientists or engineers might have copies mailed directly to their homes or offices. This seems the fastest and easiest way of keeping abreast of advances in such a busy world, where we have to work long and hard hours and have precious little time to attend courses or seminars.

Scientists and engineers have to be altruistic and unselfish, for hoarding the knowledge would certainly slacken the pace of advance in knowledge. Moreover, technology has become so complex today that no one person could be an expert in everything. It would definitely require many experts to pool their skills and knowledge together to develop a technology. The one-man inventor is now a dying breed and is being replaced by a team of bright scientific researchers each of whom might be an expert in a field of their own.

It Takes More Than Humans To Develop A Technology

Even though the technology and the brains are there, a new technology could not be developed because the tools and equipment are not available as yet. For what is a new technology, if the tools for developing it are not available, but a pipe-dream? Charles Babbage, for example, was not able to complete his mechanical computer because the tools for manufacturing some of its parts were not available.

Moreover, the scientist-engineer is also bound by limits of the law and social-political factors which could be a serious hindrance to his work. If his work involves the use of ozone-depleting chemicals or "green-destroying" products, it is unlikely that it would receive much sanction or encouragement from the authorities or his fellow-beings. If he were involved in weapons of war or destruction, the pressure on his conscience and his efforts would be all the greater. Mankind, as sane beings, should however discourage, if not ban, the development of weapons of destruction.

The Good Of All Mankind

The scientist-engineer to be an effective one should keep the welfare and interest of fellow-beings the upper-most in his mind.

It would be a sheer waste of productive time to develop a thing that no one wants or finds useful which could only be regarded as a "crazy" invention, though it might be a very clever one. It is pointless to get a patent and be proud of it, if nobody buys the idea. You could make yourself prouder and probably happier by doing other things, for example, charitable work, which would surely be appreciated.

Inventions and advances in technology must be able to better the lives of our fellow-beings. In this respect, a cure for AIDS or a pill that stops the aging process would be a dream technology come true. For those who are in search of a perfect friend, how about a talking robot that could converse intelligently (even brilliantly) with you (as well as console you) on how to solve your personal problems?

Useful Knowledge

Advances in the field of electronics seem to affect our lives more than anything else. The laser discs, the camcorders, the computers, the karaoke sound systems and many more, are all there to titillate our tastes for the higher things in life. Advanced electronics technology, for example, could now give us Virtual Reality, wherein computers could give us an imaginary world which looks real, wherein an environment could be created by computerised electronic gizmos to make it look so realistic as to be believable.

The scientist or engineer working in the research laboratory conducting researches on electronic designs or doing chemical researches could now rely on Computer-Aided-Design or CAD nowadays to bring out new models and evaluate their performances (all via the computer and its dependent software) instead of having to spend time building an actual model which may be a time-consuming and futile effort (if the model does not work out). This certainly helps to hasten research and development work, making it possible for technology to advance at a much faster rate than ever before.

The effects of technological advances are cumulative. Advance in technology not only betters the quality of our lives, it also contributes directly towards its own advancement. To function well, the scientist-engineer should have a sound knowledge of computers and software such as Prolog, C, Excel, OSS and the like. It seems that such useful knowledge is rather indispensable for many research and development activities, where experiments or models based on new designs could be "computer-created" and evaluated instead of physically created, which could be a very time-consuming and frustrating effort.

As a matter of fact, the field of computer software development or computer programming is in itself a fast advancing field. Scientists-engineers should know how to use the latest software to enhance their own work, and learning how to use them well could be a life-long learning process.

Conclusion

The advances in technology today are so great as to put us all in awe of their effects on our lives. Automated telling systems, automated ticketing systems, local area networks (LANs), and many more, all bring smooth efficiency to the way our lives are being lived.

The modern technical manager must be keenly aware of new technologies and be ready to utilise them for the sake of development and improvement. Adopting a conservative attitude towards such things would make him a loser in the technological race.

An alert, keenly aware and open mind is today an indispensable item in the race for techno-supremacy. We hear about countries wanting to import technologies and scientists and engineers being sent abroad (especially to the U.S.) for training.

To expect some modicum of success in their fields of work, technical managers and engineers must not forget to enrol themselves as students in the University of Life-Long Learning.

Labour Problem

Manufacturing has always been a labour-intensive activity. Recruiting labour for electronic assembly work has been harder and harder because of the tight labour market. Moreover, certain jobs are so monotonous that labour turnover becomes high. Furthermore, the human worker has his quirks.

Mechanisation And Automation

On the other hand, machines used in production or manufacturing activities have been playing a more and more important role. They not only perform a job more efficiently, they are super-workers when compared to humans.

The normal "conveyor line" system of electronic assembly has evolved to the automatic assembly lines or systems known as the "wave-soldering" systems, years ago. In wave-soldering, instead of having electronic components manually inserted and soldered onto the printed circuit boards by human hands, the inserting and soldering are done by the machine automatically which gives speed and accuracy resulting in comparatively little rejects. The normal active and passive components such as transistors, resistors and capacitors are still being used. The components are still soldered on only one side of the printed circuit board. Mass production is now carried out on a much larger scale, as compared to merely "hand-soldering" assembly.

The latest "in" technology in manufacturing is surface-mount assembly. This method of assembly depends on the "pick and place" robotic arm to do the job. An average surface-mount equipment could assemble 3,000 to 6,000 components an hour, with the "higher end", more expensive type capable of more than 10,000 components per hour. In this method of production, unlike the wave-soldering method, there is much more simplification and standardisation so that a wide array of components could be automatically inserted and soldered onto the printed circuit boards. Many components are packaged in the same way (thus, it is not possible to tell at a glance what kind of components they are, unless we refer to their part numbers), which is "standardisation" - this means that it is not possible to distinguish, e.g., between a diode and a resistor, without referring to their part numbers. Moreover, the components come in "tape and reel" form and are automatically fed into the surface-mount "pick and place" equipment like a reel of bullets being fed into the barrel of a machine-gun. This method of manufacturing necessitates that electronic components be specially manufactured and packaged, resulting in a new method of component manufacturing.

Moreover, the components (known as SMDs or surface-mount-devices) are now miniaturised (much smaller) and could be assembled on both sides of the printed circuit board (as compared to only one side for wave-soldering). This means that a much larger quantity of electronic components could be packed into a printed circuit board of the normal size. As consumer electronic equipment such as computers and calculators have "shrunk in size" for the convenience of the user (e.g., note-book and palm-top computers), their accompanying electronic circuitries have also to reduce in size. This is when surface-mount technology would play a useful role. This also necessitates the development of new types of soldering and desoldering tools for board-level rework or repairs, as the distance between components is now much reduced. Moreover, as the components are so much smaller, they are easily damaged when exposed to much heat (they over-heat faster), which make the selection of soldering and desoldering tools for rework/repair purposes critical as the heat from a soldering iron or desoldering iron tip could easily cause damage. This means that for some sensitive components such as integrated circuits, hot air is being used for soldering and desoldering work (in repairs or rework) instead of direct contact with a soldering or desoldering tool.

How does surface-mount work? The whole equipment comprises a robotic arm or arms, components feeders or trays and a drying oven. First, the printed circuit board is applied with a layer of paste on the side where the components are to be soldered, when it is in place and ready to receive the components. The robotic arm, which is programmable, would then pick the component from the feeder or tray and insert it on the printed circuit board, one by one. The component stays on the board firmly enough not to be dislodged as the paste keeps it in place.

After all the components or SMDs are in place, the whole board goes into an oven, which melts the paste so that the components become firmly entrenched on the printed circuit board. The method of heating in the oven is either by infra-red light or normal heat. However, of late, it has been found that the infra-red method of heating has a "light-shadow" effect problem, i.e., some components, being obstructed by the larger ones, would be out-of-reach of the infra-red light and thus would not be firmly in place.

Surface-mount is the newest in manufacturing technology, with the prices of the machine ranging from S$200,000 to S$750,000, or, more. And not all companies could afford it; certainly, it is out of the reach of many smaller companies. If for some reasons, your company wishes to go into surface-mount, you could consider the more established brands such as Dynapert, Panasonic and Fuji and the less established brands such as Technimount. There are of course many other brands. But, most companies would be concerned about the back-up technical support and the availability of spare parts provided by the supplier.

Computer-Aided-Manufacturing (CAM)

Now whole production lines could be automated and computerised, i.e., controlled and linked by computers. This is known as computer-aided-manufacturing (CAM).

Another relatively new concept in manufacturing is the flexible manufacturing system (FMS), whereby robots are well-utilised and the same production line could manufacture a variety of products. For example, a shoe manufacturing line could be converted into tooth-paste manufacturing.

With such new methods of manufacturing mass production could now proceed at an unprecedented pace, when production processes have become simplified further and the parts used have become more standardised.

This is also helped by the fact that many large circuits with many kinds of electronic components are now being "reduced to the size of the thumb" through advanced semi-conductor technology, which gives us many advanced ICs (integrated circuits). This obviously helps reduce the number of manufacturing processes (process simplification).

Modern Test Equipment

The normal method of testing assembled printed circuit boards is by getting a technician to use the multimeter, logic probe or oscilloscope.

However, with modern technology this may not be necessary. Modern in-circuit testers could accommodate the whole board, test the circuit by injecting voltages and signals, and give a computer print-out of the results, giving a mention of the defective sections as well.

Before the advent of computer-aided-design (CAD), an electronic designer has to bread-board assemble his just-designed circuit for testing. But CAD has made all this redundant. The designer could simulate and test the circuit he has designed with the computer only (using such software as HSPICE, i.e., he does not have to build a model to test based on his design. (What a saving of time!)

For circuits-on-a-chip (integrated circuits), universal programmers are available which allow one to program and test them. PALs, EPLDs, EEPLDs, GALs, DRAMs, SRAMs, microcontrollers, EPROMs and EEPROMs could be programmed and tested by a programmer about the size of a book. The "higher end" programmers could program up to about 32 ICs at one go very quickly and efficiently and could perform this job on up to about 3,000 ICs a day (say, eight hours) while the "lower end" ones could provide the same service for about 300 to 500 ICs a day (eight hours).

Moreover, many half-finished and fully-finished products have to be tested thermally as well as with humidity. The modern temperature test chamber is able to provide temperature cycles for "thermal shock" tests of parts and finished products, as well as for humidity tests. Such tests are important for equipment which could not afford to fail under varying weather conditions, e.g., in the case of automotive parts, wherein the car has to "bear" all kinds of weather condition - hot, cold, dry and wet weather conditions. The modern temperature test chamber is normally programmable and could be linked externally (optional) to a computer, through the RS 232 or GP1B interface card, for example. There are even special ovens for testing integrated circuits known as "burn-in" ovens.

Advances

The amount of equipment available for manufacturing is so amazing that the scope of this book would not be able to do enough justice to them.

As technology is still evolving, it is difficult to tell whether the equipment now used in manufacturing would be as useful in five to ten years' time.

The modern technical manager or engineering manager, whose main task is to support the production department, has to be keenly aware of the advances in manufacturing engineering.

CONCLUSION

Engineering, especially electronics engineering, is a fast-moving field. The engineering manager and his team have to keep abreast of the latest developments in technology. This makes their work challenging and refreshing. They have to look around and try to move forward all the time. They have to know what the latest technologies used by competitors are.

It pays for them to read widely the technical journals that are available. It also pays for them to attend courses and seminars to upgrade their know-how.

This book will somewhat enable the reader to improve on his technical, as well as managerial, know-how.

People are complex beings with complex needs, e.g., need for security, esteem, companionship, self-actualisation and physical preservation (food, clothing and shelter). Moreover, humans are emotional, capable of jealousy and envy and have likes, dislikes, preferences and prejudices. In any organisation, it is the manager's difficult responsibility to manage people, many of whom might be difficult to manage or even unmanageable, especially those with the penchant or tendency to criticise, rebel, stir trouble by gossiping or circulating rumours, or sabotage their peers and/or their management. The responsibilties of a manager have been touched on in the author's related book BUSINESS ADMINSTRATION MADE EASY AND PRACTICAL and they include tasks such as planning, organising and communicating. Knowing the responsibilities of managing is of course elementary and basic. But much more important for successful management or managerial results would be the proper execution of management tasks.

How should a manager manage people? This is a very important question which deserves very careful consideration. A manager who manages machines would find the task relatively pain sailing as machines do not talk back or argue, do not need tea or toilet breaks and would "obey" the manager faithfully. Not so for the human machine, who needs tea, lunch, dinner and even supper breaks, as well as toilet breaks and sick leave when ill, and rest when fatigued. For the clueless, the inexperienced and the fool, managing is most easily carried out by firmly issuing a set of instructions or orders using one's managerial authority, closely monitoring and checking the work of the subordinates and utilising threat of disciplinary action to get the work done well and completed on time. This would inevitably result in stress and tension, fear and unhappiness in the work-force. The mantra here is evidently "if you do not do your job well, you would be disciplined and even fired". This is work-place dictatorship or feudalism. Our modern society has practically done away with authoritarianism, dictatorship and monarchy, having regarded them as scourges, and practises democracy which is evidently considered the ideal society. But sadly most of us have been living in a "fake" democracy, for most of us spend our working hours in the work-place where democracy is hardly ever practised, for employees and workers are subject to the authoritarianism of their employers and their representatives, the managers. We all know that in the companies we work for for most of our waking hours there is no true democracy and the employers and their representatives, the managers, arbitrarily call the shots - they could dictate that you wear their uniforms, they dictate that you report to work and leave at certain times, that you could not do certain things at certain times, that you have to behave in a certain manner, in short to be disciplined, that you cannot disobey your managers which is insubordination and liable for disciplinary action, that you would be paid so much as you are only worth so much for the work you are doing, that you shall not do this, this and that, etc.. As an employee, you are more or less a slave, with the employer and his representatives, the managers, being the slave-drivers. Apparently, the rationale for all this is that in the work-place for the sake of productivity, profitability and everyone's economic survival, as well as safety, even welfare, a certain degree of discipline is necessary. We read reports in the media that people nowadays work longer hours but are producing less than before. What is happening? We have not been improving. We do not appear to be working smarter. The author suspects that the managers, the representatives of the employers, or even the employers themselves if they are also the company's managers, have not done a proper job. They could have introduced incorrect work procedures and methods and/or they could have failed to motivate their staff or could have even de-motivated their staff.

To be effective, the manager evidently needs to win over the hearts and minds of his subordinates so that they would go the extra mile for him and the company, he being the representative of the company. How could the manager achieve this most important objective? There cannot be any hard and fast rules for achieving this. It would depend on the kind of subordinates the manager has. He needs to understand their personalities, their needs, their strengths and their weaknesses. In fact, to begin with, the manager in order to get things done and achieve the desired results must have the right kind of people with the right skills working for him. Recruiting the right people is actually the crucial first step. Whoever does the recruiting, preferably the manager himself, has to be a good judge of people. If his judgement is poor and he hires the wrong people and has to fire them later, these firings would be likely to adversely affect the morale of the work-place, for the erring manager who fires the staff would be likely to be perceived as a ruthless and bad person who is not worth working for. Hence, recruiting is crucial and should be very carefully carried out. For the manager to have one or two bad judgements on hiring is already quite bad. If a manager continually makes many such wrong recruitments, then it is evident that he is poor judge of people and/or has not carried out his recruitment task properly or responsibly. This manager should account for this faux pas and should be fired if there is no good reason for his failure to hire the right persons for the job. However, it should be borne in mind that certain skilled personnel might not be available at that point of time when the company needed them and the manager might just have to make do with the best which were available then, in which case the manager could be excused for not having hired the right people. It seems at odds that workers are always fired for being poor workers while the recruiters are scot-free. It is more rational that the recruiter be fired together with the wrong hire if it is clear that the recruiter had messed up the hiring job one too many times due to his poor people judgement and/or lack of care and responsibility.

As stated above, people are complex and have complex needs. Therefore, to manage people well first of all involves really understanding them as individuals as well as fully understanding their needs. Many managers shout, scold, even threaten with disciplinary action, and carry out all kinds of disrespectful activities, when dealing with staff. This must be a no-no. The basic tenet in manager-staff relationship has to be one of mutual respect. If the manager expects the staff to respect him he must first of all show respect towards them as fellow-human beings. Everyone living together in this world, whether humans or animals, should live together peacefully and with respect for each other's feelings and rights, if anarchy were to be avoided and if living were to be worthwhile and dignified. However, this ideal situation does not seem to be present much of the times, making living on earth seems more like living in hell. Therefore, a manager should, very importantly, first of all learn and try to be a good psychologist, a person who understands well how others think and feel, in order to manage them well and get the best possible results through them. He should understand that the staff working with him represent him and are performing some tasks for him and he would be judged by his management on how well he could get results through his staff. In short, he should be able to strongly motivate his staff. A manager can never motivate his staff by carrying out disrespectful activities towards them such as scolding them and threatening them with disciplinary action. He should be able to persuade them, to win over their hearts and minds, to encourage them, to be able to make them willingly, and not grudgingly, go the extra mile for him, perhaps even die for him. Needless to say, the manager should be protective of his staff and have their welfare at heart, which are all important aspects of motivation. To be a really great manager, he should be a great leader of people and should be able to lead by example, e.g., by being hands-on. Here, the author would like to describe his hero and his model for management and leadership. This model is Field-Marshal Erwin Rommel, a charismatic tank commander in Hitler's armed forces during World War Two. This brilliant general had been an icon in his homeland and to his enemies. He was brilliant in the deployment of battlefield tactics. He was worshipped by his men, having had great respect and welfare for the men under him. Unlike other generals who merely barked orders somewhere behind the frontline, he would be really hands-on and would be at the very frontline itself, in fact, in front of his men, leading his men in charging at the enemy. He was evidently not afraid of death when he displayed such great courage and had narrow escapes himself. He was like a God to his men, who all really adored him and were evidently willing to die for him. So great was he as a motivator and leader of men. Even his enemies respected him greatly for his bravery and exemplary leadership. He was also much respected as a cunning and brilliant tactician in battle. As a manager and leader of men, we should all be like Field-Marshal Erwin Rommel, who had led his men to great success in battle by example. It would be good for managers to read about the escapades of this very great German general and thereby, hopefully, learn leadership and man-management skills from this great man.

Managers hopefully would feel inspired in leading their staff with a kind of missionary zeal. Of course, to be so inspired, the manager has to have great belief in the nobility and importance of his job, e.g., the achievement of the manager's objectives benefits the society at large. The manager, to be a great one, should thus be doing work which he strongly believes in. Unfortunately, for many companies, and for the society at large, many managers who have studied management in school do not apply what they have learnt but have become bullies of men and brutes, abusing the authority they are invested with, becoming dictators to their men, doing all the no-no's of management, contradicting all the textbook knowledge they were supposed to have acquired. Perhaps, they had studied management techniques just for the piece of precious paper, the diploma or degree.

As a parting shot, the author would like to suggest that the manager leads his team like a successful soccer captain leading his team of players, assigning roles to his staff according to their strengths and ensuring that his team, including himself, works together as a well-oiled and well-coordinated whole, with good teamwork. Managers therefore have much to learn from a successful soccer team.

The author would like to exhort all managers to note carefully and ponder what has been said so far.

However, technical managers need more than good people management skills. They should also be able to manage the technology being utilised. This means that they should be technically inclined, keen to learn new techniques or technologies and able to understand and use them well (e.g., by reading up, attending talks and courses or simply picking the brains of the experts in the respective technical areas). In other words, the technical manager should be able to master the requisite technologies and guide his subordinates in the use of these technologies. Managing technical people is hence a more difficult, more challenging task. In order to have the respect of his technical staff, the technical manager should have sufficient technical competence, which could be a problem as technology is arcane, not easily accessible to the uninitiated, may take a long time, years, to master, and, worse, changes rather fast, perhaps faster than the technical manager's rate of mastery. In the end, the technical aspect of the technical manager's job may be tougher than the people management aspect. Whenever the technical staff could not solve some technical problems, they would normally approach the technical manager for help in finding the solutions. If the technical manager were not able to help solve these technical problems, his staff might lose confidence in and respect for him, which would make it difficult for him to manage them. Technical managers should therefore make themselves well-versed with the technologies they are managing, thereby gaining the confidence of their staff and being able to earn their respect and cooperation. There is in fact no faster, better way to learn a technology than by being hands-on, by doing, by being directly involved, by getting the hands dirty. The technical manager, by being hands-on, would be leading by example, like the above-mentioned brilliant German general Erwin Rommel, instead of just barking orders. He should also rely on experience, common sense and judgement when performing his tasks.

APPENDICES:
EXHIBITS OF ENGINEERING AND OTHER IMPORTANT DOCUMENTS

APPENDIX 1

MODEL INTRODUCTION CHECK-LIST

1. Schedule determination - determine time of mass production.
2. Signal source - determine what equipment to buy if necessary.
3. Alignment jigs/operation tester?
4. Cabinet moulds?
5. Application for safety approval? (EMI, UL, SAFETY, DHHS, etc..)
6. Headquarters Pre-production.
7. Sample sets from Headquarters?
8. Drawings from Headquarters?
9. Parts list from Headquarters?
10. Pre-production.
11. Component kits from Headquarters for free production?
12. Labels/Carton Boxes - Drawings to Singapore suppliers & approval of samples (Engineering/QC).
13. Cushion mould/metal brackets - Drawings to Singapore suppliers & approval of samples (Engineering/QC) - Get list first?
14. PCB & other plastic parts - Issue drawings to Singapore suppliers & approval of samples (Engineering/QC) - Get list first?
15. Order auto-insertion holder jig?
16. Issue parts lists to Materials/Purchasing?
17. Headquarters to issue Specification Manual to co.?
18. Issue parts lists & PWB main to sister factory for auto-insertion?
19. Issue parts list & PWB main to main factory for manual insertion?
20. Pre-production run.
21. Actual PWB main Alignment Test/Operation Test?
22. Actual Sub-assembly for Pre-production Run.
23. Actual Final assembly for Pre-production Run.
24. Completion of measurements against specifications.
25. Compile & issue process drawings to production (instruments).
26. Compile & issue jig list.
27. Fabricate & issue production jigs, including pre-setter, indicator jigs.
28. Vibration/Drop tests (Engineering & QC).
29. Material costing.
30. Get TS from Headquarters.
31. Breakdown of TS for various production sections - Total TS.
32. Introduce model to production/QC.
33. Give sample sub-assembly & complete set to production/QC.

34. Necessary to send engineer to Headquarters to study pre-production?
35. Issue Incoming QC new material list for information.
36. Requirement for new equipment/tools?
37. Is new line layout necessary? Advise production.
38. Comparison list(s)?
39. Production Flow Chart?
40. Distribution of sample sets to production, QC drop test. Chassis rework technicians, engineering samples, Headquarters, etc..
41. Assign engineer to line when production starts.
42. Issue bar code numbers.
43. Engineering report on problems encountered in Pre-production run.
44. Issue & keep track of all ECNs.
45. Receipt of all drawings.
46. Cost study.
47. Performance Check & Reports for Pre-production sets.
48. Solve all technical problems on line as quickly as possible.
49. R & D.

APPENDIX 2

File No.: _____

TO: ENGINEERING

PREPARED BY:

APPROVED BY:

DATE: 21/7/86

TECHNICAL INFORMATION REPORT

TOPIC: NEW MONOCHROME MONITOR, MODELS 6231, 6489, 6512 & 6634

CONTENTS

Attached herewith are the data for the performance checking & reliability testing for the above models.

The performance can't be confirmed as the XEROX PC we have has broken down and we are still waiting for the new computers to arrive.

Preliminary problems encountered were top curl, focussing & vertical jitter. These, we must confirm with the computer first.

Also, certain readings such as picture size & centering cannot be confirmed without the computer.

For the performance check, the Olivetti PC and the National signal generator are used instead.

All the readings are within specs, except for the rise & fall time which we must confirm with the PC itself.

REPLY REQD.: YES/NO BY

APPENDIX 3

PERFORMANCE CHECKLIST FOR MONOCHROME MONITOR

MODEL:

DATE: CHECK SHEET NO.: 1

	CHECK ITEM	SIGNAL	CONDITION			RESULT		SPEC
			BRT	CONT	OTHERS			
1	POWER							
a	LOWER LIMIT	WHITE	MAX	MAX		98V		98V
b	UPPER LIMIT	WHITE	MAX	MAX		139V	OK	127V
c	CONSUMPTION	WHITE	MAX	MAX	159 NITS	43.5W	OK	
				MAX	137 NITS	42W	OK	55W
2	B+							
a	SETTING	H	MAX	MAX	137NITS	11.87V	OK	121 0.2V
b	CURRENT	H		MAX	159NITS	1.96A	OK	
3	ANODE VOLTAGE							
a	AT 0 BEAM 100 UA					12.7KV		
						12.1KV		
b	REGULATION					0.6KV	OK	0.6KV
4	HORIZONTAL SYNCHRONIZATION							
a	SETTING					25.76KHz		25.86KHz
b	PULL-IN RANGE					25.26 - 26.37KHz	OK	0.5KHz
c	HOLDING RANGE					24.65 - 27.16KHz	OK	0.5KHz
5	VERTICAL SYNCHRONIZATION							
a	SETTING							
b	PULL-IN CHARGE					37-59KHz	OK	10-60KHz
c	HOLDING					34-61KHz	OK	10-60KHz
6	PICTURE QUALITY							
a	RESOLUTION							
b	OVERSHOOT			MAX	86 NITS	6%	OK	
c	SMEAR						OK	
d	FOCUS	@ & H				LEFT SIDE DOOR		
e	VIDEO SATURATION						OK	
f	BENDING ON TOP	H				TOP CURL		
g	VERTICAL JITTER	WHITE				SLIGHT JITTER		
h	SHAKING OF PIX							
	i) DUE TO POWER RIPPLE	H					OK	
	ii) DUE TO MAGNETIC FLUX	H		MAX	86 NITS		OK	
i	UNIFORMITY OF BRIGHTNESS	WHITE				23 76 86 80 74	AVERAGE 79.8 NITS	

53

7	STARTING & TURNING OFF CHARACTERISTICS							
a	BRIGHT SPOT	WHITE				NIL	OK	
b	SYNCHRONIZATION	WHITE				NIL	OK	
c	DISAPPEARANCE OF RASTER							
d	DISAPPEARANCE OF RASTER	WHITE				NIL	OK	
e	WARMING UP DRIER	WHITE				NIL	OK	
8	BRIGHTNESS	WHITE		MAX		EXT	OK	34 NITS
	NATIONAL S/G					BRT RANGE		TO
a	SETTING LEVEL	GRAY		MAX		= 22 NITS		137 NITS
		SCALE				TO 162		
						NITS		
						86 NITS	1ST LEVEL	86 NITS
							2ND LEVEL	46 NITS
							3RD LEVEL	22 NITS
							4TH LEVEL	0 NITS
				MAX	MAX		1ST LEVEL	137 NITS
9	HORIZONTAL WIDTH i) SETTING ii) MAX iii) MIN							

APPENDIX 4

PERFORMANCE CHECKLIST FOR MONOCHROME MONITOR

MODEL:

DATE: CHECK SHEET NO.: 2

	CHECK ITEM	SIGNAL	CONDITION			RESULT		SPEC
			BRT	CONT	OTHERS			
1	POWER							
a	LOWER LIMIT	WHITE	MAX	MAX		196V	OK	203V
b	UPPER LIMIT	WHITE	MAX	MAX		263V	OK	254.4V
c	CONSUMPTION	WHITE	MAX	MAX	137 NITS	49.5W	OK	100W
				MAX		48W		
2	B+							
a	SETTING	H		MAX	137 NITS	12V	OK	12+0.2V
b	CURRENT	H		MAX	137 NITS	1.95A		
		WHITE	MAX	MAX	165 NITS	2.03A		
3	ANODE VOLTAGE							
a	AT 0 BEAM 10 MA					12.75 KV	OK	12.2KV
b	REGULATION					0.55KV	OK	0.6KV
4	HORIZONTAL SYNCHRONIZATION							

a	SETTING					25.7KHz		25.86KHz
b	PULL-IN RANGE					25.27 - 26.38KHz	OK	0.5KHz
c	HOLDING RANGE					263V 24.78 - 27.17KHz	OK	0.5KHz
5	VERTICAL SYNCHRONIZATION							
a	SETTING					44Hz		
b	PULL-IN CHARGE					38-59Hz	OK	10-60KHz
c	HOLDING					35-62Hz	OK	10-60KHz
6	PICTURE QUALITY							
a	RESOLUTION							
b	OVERSHOOT			MAX	86 NITS	6%		10%
c	SMEAR						OK	
d	FOCUS	@ & H				LEFT SIDE		
e	VIDEO SATURATION						OK	
f	BENDING ON TOP	H				TOP CURL		
g	VERTICAL JITTER	WHITE				SLIGHT JITTER		
h	SHAKING Of							
	I) DUE TO RIPPLE	H					OK	
	ii) DUE TO MAGNETIC FLUX	H					OK	
i	UNIFORMITY OF BRIGHTNESS	WHITE				82 78 86 76	AVERAGE 80.5 NITS	
7	STARTING & TURNING OFF CHARACTERISTICS							
a	BRIGHT SPOT	WHITE				NIL	OK	
b	SYNCHRONIZATION	WHITE					OK	
c	DISAPPEARANCE OF RASTER							
d	DISAPPEARANCE OF RASTER	WHITE				NIL	OK	
e	WARMING UP DRIER	WHITE				NIL	OK	
8	BRIGHTNESS OLIVETTI		MAX	MAX		160 NITS	OK	137 NITS
a	SETTING LEVEL			MAX	86 NITS	1ST LEVEL 2ND LEVEL 3RD LEVEL 4TH LEVEL	86 NITS 42 NITS 22 NITS 0 NITS	86 46 22 0 } OK
9	HORIZONTAL WIDTH i) SETTING ii) MAX iii) MIN							
10	LINEARITY							
a	VERTICAL	+				7%	OK	10%
b	HORIZONTAL	+				8%	OK	10%

11	CONTROL ABILITY OF EXTERNAL CONTROLS							
	i) BRT VR			MAX	FULL RANGE		OK	
	ii) CONT VR		MAX		FULL		OK	
12	SIZE REGULATION	WHITE		MAX	10-40 FI-L	H:0.5% V:1.2%	OK OK	2% 2%
13	ARISE TIME (tr)	WHITE		MAX		32ns		
14	FALL TIME (tr)	WHITE		MAX		45ns		

APPENDIX 5

OUTGOING SAMPLING METHOD

B.1 SAMPLING PLAN

B.2 INSPECTION STANDARD

B.3 CLASSIFICATION OF LOTS

B.4 CLASSIFICATION OF DEFECTS

B.5 AQL JUDGEMENT

B.6 SAMPLING PROCEDURES

B.7 TYPES OF INSPECTION

B.8 PATROL INSPECTION

B.1 SAMPLING PLAN

Single, normal sampling inspection by attributes according to MIL STD 105D.

Applied only to Process and Warehouse Inspection.

B.2 INSPECTION STANDARD

B.2.1   Inspection Level

1.   Level II: Applies to Manufacturing Process Inspection and Electrical Performance Inspection.

2.   Level S-3: Applies to Internal Inspection and Packing Inspection.

For details on the inspection conducted, please refer to paragraph B.7: Types of Inspection.

B.2.2   Switching Procedures

Initially, NORMAL INSPECTION PLAN is used. After that, switching procedures may be followed.

NORMAL --→ TIGHTENED	2 lots of 5 continuous lots rejected
TIGHTENED --→ NORMAL	5 continuous lots accepted

NORMAL --→ REDUCED	10 continuous lots accepted and defective Qty.. Limit Numbers for reduced inspection.
REDUCED --→ NORMAL	1 lot rejected or production irregular or circumstantial consideration

B.3 CLASSIFICATION OF LOTS

B.3.1   Definition of Lot

A lot is composed of the parts or products assembled on the same production line, on the same day, and of the same model.

B.3.2   Categorization

1.   Sub-Assembly/Chassis-Alignment Stages:

One lot - total number of work-in-progress units on one trolley or within one bin.

2.   Monitor/Color TV Stages:

One lot = quantity manufactured during the specified time interval of 10.00 am to 3.30 pm (called Morning Lot) or 3.30 pm to 10 am (called Afternoon Lot).

B.4 CLASSIFICATION OF DEFECTS

B.4.1   Definition of Defect

A defect is a deviation from specified process, drawings, or specifications and includes those caused by ordinary transportation, application conditions, and handling.

B.4.2   Categorization

All defects are categorized as critical-major, major, or minor defects.

1.   Critical Defect:

Defects which cause hazards to human.

2.   Major Defect:

A deviation which causes the main performance and function of the product not to be achieved or not to be stable. Also those defects which significantly impair the commercial value, and reliability of the product.

3.   Minor Defect:

A defect which deviates slightly from the given specifications and standards. Also those defects which only slightly reduce the commercial value and reliability of the product.

B.5 AQL JUDGEMENT

B.5.1   Defect Judgement Criteria

1.   For General Inspection Level II

Judgement criteria for all defects are as follows:

Critical Defect: AQL 0%

Major Defect: AQL 1.0%

Minor Defect: AQL 2.5%

2.   For Special Inspection Level S-3 Judgement criteria are as follows:

Critical Defect: AQL 0%

Major or Minor Defect: AQL 2.5%

B.6 SAMPLING PROCEDURES

Sample size for inspection is based on that of the Sampling Table. Samples are taken RANDOMLY from the lot.

For details on treatment for accepted/rejected lot and accepted/rejected single unit, please refer to Outgoing Procedures for Disposition of Non-Conformance Items.

B.7 TYPES OF INSPECTION

B.7.1   Process Inspection

1.   Manufacturing Process Inspection

Sampling size based on General Inspection Level II, Sampling Table.

Inspection areas include mechanical, electrical, constructional, cosmetics, soldering, mounting, wiring connection, and wire dressing.

2.   Critical Component Inspection

Sub-Assembly Stage: All QC samples with critical components are checked for their part number (indication) and mounting.

Chassis-Alignment Stage: One unit per trolley-lot is checked.

B.7.2   Warehouse Inspection

1.   Performance Inspection

Sampling size based on General Inspection Level II, Sampling Table.

Inspection includes checking display quality, display geometries, shock test, cosmetics, and other functional checks.

2.   Internal Inspection

Sampling size based on Special Inspection Level S-3, Sampling Table.

Inspection includes checking approved indications, component mounting, soldering condition, and required labels.

3.   Packing Inspection

Sampling size based on Special Inspection Level S-3, Sampling Table.

Inspection includes checking packing condition, presence of accessories, and control setting condition.

4.   Electrical Tests

Hi-Pot test and grounding test are carried out twice a day; once in the morning and once in the afternoon. These tests are conducted for safety control reasons.

B.7.3   Technical Tests

1.   Safety Tests

TEST ITEM	SAMPLING SIZE
1. Power Consumption	One unit/shipment lot
2. Leakage Current	One unit/shipment lot
3. Strength Relief	One unit/shipment lot
4. Insulation Resistance	One unit/shipment lot
5. Critical Component Check	One unit/shipment lot
6. Hi-Pot/Grounding Tests	One unit/shipment lot
7. HV/X-Radiation	One unit/day/line
8. Wire wrapping	Once per week
9. FCC Measurement: One set sent to HQ yearly	One set sent to HQ yearly

2. Manufacturing Control Tests

TEST ITEM	SAMPLING SIZE
1. Solder Bath Temperature	Once daily
2. Specific Gravity of Flux	Once daily
3. Magnetic Champer Calibration	Once daily
4. Performance Measurement	Special Inspection S-2, per shipment lot
5. Daily Electrical Measurement	Special Inspection S-2, per shipment lot

3. Reliability Tests

Please refer to Section 6 of this Quality Plan for details.

B.8 PATROL INSPECTION

1. Types of Inspection

Patrol inspection on Process and patrol inspection on Product shall be carried out to verify that the manufacturing instructions are followed and the production process is under control.

2. Frequency of Inspection

Each type of patrol inspection shall be conducted twice monthly.

OUTGOING INSPECTION MANUALS AND SPECIFICATIONS

C.1 PROCESS INSPECTION MANUALS

C.2 WAREHOUSE INSPECTION MANUALS

C.3 TECHNICAL MEASUREMENT MANUALS

C.4 PROCESS INSPECTION SPECIFICATIONS

C.5 WAREHOUSE INSPECTION SPECIFICATIONS

C.6 TECHNICAL MEASUREMENT SPECIFICATIONS

C.1 PROCESS INSPECTION MANUALS

Tentatively, the following Process Inspection Manuals shall be prepared:

1. Inspection manual for PWB MAIN/PWB SOCKET
2. Manual for PWB VOLUME
3. Manual for PWB POWER/POWER TRANSFORMER
4. Manual for EARTH WIRE
5. Manual for DEFLECTION YOKE

C.2 WAREHOUSE INSPECTION MANUALS

Tentatively, the following Warehouse Inspection Manuals shall be prepared:

1. Inspection manual for Performance Inspection
2. Manual for Electrical Tests (Hi-Pot/Grounding)

C.3 TECHNICAL MEASUREMENT MANUALS

C.3.1 Safety Test Manual

The following manuals shall be prepared:

1. Power Consumption Test Manual

      2.   Leakage Current Manual

      3.   HV/X-Radiation Manual

      4.   Strength Relief Manual

      5.   Insulation Resistance Manual

      6.   Wire Wrapping Manual

C.3.2   Reliability Tests Manual

The following manuals shall be prepared:

1. Life Test Manual
2. Vibration Test Manual
3. Drop Test Manual
4. Humidity/Temperature Test Manual

C.3.3   Manufacturing Control Manuals

The following manuals shall be prepared:

1. Performance Geometrics Measurement Manual
2. Magnetic Field Measurement Manual

C.4 PROCESS INSPECTION SPECIFICATIONS

Process inspection specifications for mechanical, electrical, constructional, cosmetics, soldering, mounting, and wiring judgement shall be compiled.

C.5 WAREHOUSE INSPECTION SPECIFICATIONS

Warehouse inspection specifications for visual, mechanical, functional, internal, and packing inspection shall be compiled.

C.6 TECHNICAL MEASUREMENT SPECIFICATIONS

Specifications for Safety Measurements, Reliability Measurements, and Manufacturing Control Measurements shall be compiled.

OUTGOING INSPECTION REPORTS AND RECORDS

D.1 PROCESS INSPECTION REPORTS AND RECORDS

D.2 WAREHOUSE INSPECTION REPORTS AND RECORDS

D.3 TECHNICAL MEASUREMENT REPORTS AND RECORDS

D.4 MISCELLANEOUS ITEM INSPECTION REPORTS AND RECORDS

D.5 MISCELLANEOUS MANAGEMENT QUALITY REPORTS

D.1 PROCESS INSPECTION REPORTS AND RECORDS

The following documents are available:

1. Process Control Inspection Report
2. Checking Report for Indication of Approval Components
3. Warning Notice
4. Rework Report
5. Defective Report

D.2 WAREHOUSE INSPECTION REPORTS AND RECORDS

The following documents are available:

1. Outgoing QC Inspection Daily Report
2. Equipment Check Record Sheet
3. Internal Inspection Report
4. Finished Product Packing Inspection Report
5. Passed lot notice
6. Rejected lot notice
7. Pending lot notice
8. Warning Notice
9. Rework Voucher

D.3 TECHNICAL MEASUREMENT REPORTS AND RECORDS

D.3.1   Safety Documents

The following documents are available:

1. Wire Wrapper Maintenance Check Sheet
2. Leakage Current Test Record
3. Power Consumption Record
4. X-Radiation Measurement Record
5. Critical Component Check List
6. High Potential and Polarity Check Record
7. Strength Relief Test Record
8. Insulation Resistance Test Record

D.3.2   Reliability Documents

The following documents are available:

1. Life Test Inspection Report
2. Vibration and Drop Test Report
3. Humidity and Temperature Test Record

D.3.3   Manufacturing Control Documents

The following documents are available:

1. Specific Gravity of Flux Record
2. Solder Bath Temperature Record
3. Daily Electrical Measurement Record
4. Performance Geometric Measurement Record
5. Magnetic Field Calibration Record

D.4 MISCELLANEOUS ITEM INSPECTION REPORTS AND RECORDS

The following documents are available:

1. Patrol Inspection Report: Product
2. Patrol Inspection Report: Process
3. Defective Investigation Report
4. Technical Information Report

D.5 MISCELLANEOUS MANAGEMENT QUALITY REPORTS

The following documents are available:

1. QC Daily Report to Production
2. Final Production Summary Report: Operator Situation
3. Technical Summary Report
4. Lot Confirmation Sheet for Final Lines
5. Daily Quality Report (Graph)
6. Production Quality Monthly Report (Graph)
7. Outgoing QA Section Monthly Summary Report
8. Outgoing QA Section Annual Summary Report

OUTGOING PROCEDURES FOR DISPOSITION OF NON-CONFORMANCE ITEMS

E.1 TREATMENT OF LOT

E.2 TREATMENT OF SINGLE UNIT

E.1 TREATMENT OF LOT

E.1.1 Procedures for Rejected Lot

1. Basis for lot rejection

   a) Deviations from specifications

   b) Differences compared with engineering sample

   c) Managerial decision in extraordinary circumstances

   d) Application of AQL Accept-Reject Criteria

2. Reject Indication on Lot

   a) Sub-Assembly/Chassis-Alignment Stage A Red rectangular stamp on the Defective Report indicates to the Production Section concerned that the lot has been rejected by QA Department.

   b) Monitor/CTV Stage

   A red Rejected lot notice is attached to each row of the defective lot.

3. Formal Notification

   A Rework Voucher is issued to the Production Section concerned whenever a lot is rejected. It is then returned to QA Department after the form has been duly completed.

4. Rework and Re-inspection

   Rejected lots are reworked by the Production Section concerned and re-inspection by QA Department.

5. Involvement by Other Departments

   When the nature of the defective symptom requires investigation involving the co-operation of other departments, a meeting is called to resolve the problem.

   A Memorandum is then issued by the investigators to departments concerned.

E.1.2   Procedures for Accepted Lot

         1.   Approval Indication on Lot

             a)   Sub-Assembly/Chassis-Alignment Stage

A blue triangular stamp on the Defective Report indicates that the particular lot has been accepted by QC and may continue to the next manufacturing operation or process.

             b)   Monitor/CTV Stage

A blue Passed lot notice is attached to each row in the lot.

E.2 TREATMENT OF SINGLE UNIT

E.2.1   Procedures for Rejected Single Unit

         1.   Rejection Indication

             a)   Sub-Assembly/Chassis-Alignment Stage Rejected single unit is given to the Production Line Leader concerned for immediate repair. The repaired unit is then inspected once again.

             b)   Monitor/CTV Stage

Rejected complete single unit is attached with the red Defective Report. After Production has repaired the unit the Defective Report is returned to QC together with the repaired unit for re-inspection. The serial number of the unit is recorded.

         2.   Formal Notification

Based on the AQL judgement criteria, a Warning Notice is issued to the Production Section concerned as appropriate.

E.2.2   Procedures for Accepted Single Unit

         1.   Approval Indication

             a)   Sub-Assembly/Chassis-Alignment Stage

An alphanumeric stamp is made on those units that have passed QC Inspection.

             b)   Monitor/CTV Stage

A blue circular stamp is made on the carton box or History card of the unit that has passed QC Inspection.

APPENDIX 6

OUTGOING PROCEDURES FOR HANDLING CUSTOMER COMPLAINTS

1. ORGANISATIONAL PROCEDURES FOR HANDLING CUSTOMER COMPLAINTS
2. PRODUCT TRACEABILITY

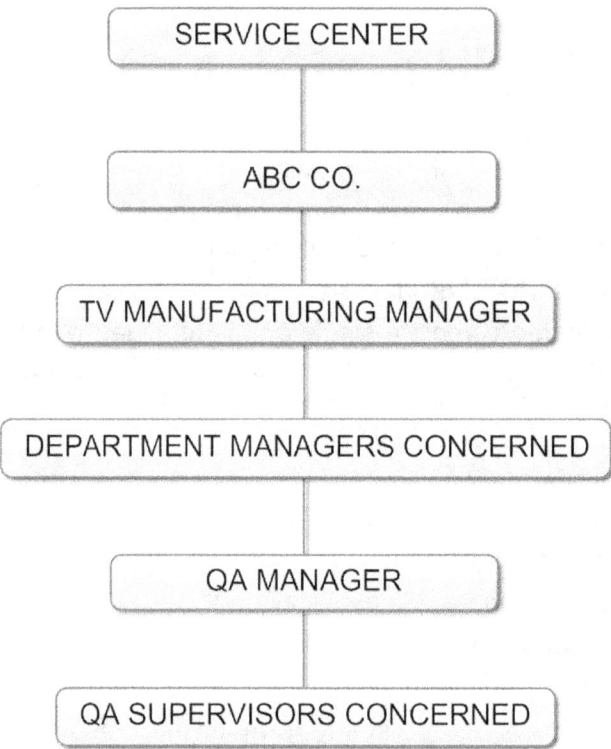

APPENDIX 7

APPROVAL OF COMPONENTS

1. **SCOPE**

    This regulation should be applied to the components for approval so as to be able to maintain the specified quality level of XYZ Products.

2. **PROCEDURE OF COMPONENT APPROVAL**

    (1) Inquiry of material for approval should be submitted to Overseas Corporation Department of Johnston-Kingston Works with the following necessary materials:-

    a) Sample to be submitted
    b) Specification for Approval
    c) Material List for the component
    d) Reliability assurance data
    e) Quality assurance system of the manufacturer

(2) Overseas Corporation Department shall issue this inquiry to the Overseas Technical Department for evaluation.

(3) After confirming the necessary materials, Overseas Technical Department shall submit the materials to Q.C. Department for evaluation following the new Component Approval Regulation to examine the Quality and Reliability of it.

(4) Having finished examination, Q.C. Department will then report the results to Overseas Technical Department with their judgement.

(5) Overseas Technical Department shall make the final conclusion with consideration of the application technique which will be applicable for all models and their safety.

After that, the list of Local Procured Approved Components shall be issued out to Overseas Corporation Department through Standards Department.

(6) Overseas Technical Department shall inform our company about the approval by forwarding us the list of Local Procured Approved Components.

(7) Applying schedule of local procured components, the localized plan of components shall be decided through thorough discussions with Overseas Corporation Department, Overseas Technical Department and our company.

3. RANGE OF APPROVAL AND FOLLOW-UP SYSTEM OF LOCAL APPROVED LIST

1) Range Of Approval

We shall apply this regulation to the following situations:

a) New component, new material, new assembled unit procured from outside.

b) Changing of material, construction of the component and manufacturing process which may result in difference of performance.

c) Changing of contractor (supplier) for component concerned.

2) Follow-up System Of Approved Components List

Follow-up of the list which have been issued out shall be the responsibility of our company. Regulation of Revision for Drawing and revised list shall be issued out together with drawing revision information.

The following components with the required quantity to be sent to Johnston-Kingston Works for evaluation:-

DESCRIPTION OF COMPONENTS	REQUIRED QUANTITY
Semi-conductors (Diode, Transistor IC)	50 Pieces
Resistor	100 Pieces
R-Cement wire	50 Pieces
Variable Resistor	30 Pieces
Ceramic Capacitors	100 Pieces
Electrolytic Capacitors	10 Pieces
IPT, Coils	30 Pieces
Trans-Power	10 Pieces
Auto-Degausing Coil	5 Pieces
Tuners (VHF & UHF)	10 Pieces
Speakers	5 Pieces
Mechanical parts (connector, power-jack)	10 Pieces
All plastic parts (Back-cover, Front panel)	3 Pieces

Metal parts (Tuner, VR, CRT brackets)	5 Pieces
Switches	10 Pieces
Lamp	50 Pieces
Fuse	50 Pieces
Delay Line	50 Pieces

APPENDIX 8

GENERAL CHECKING PROCEDURE OF ALL INCOMING GOODS

(FLOW CHART REF. NO. ICQA/023/85)

1) Goods received by our store from supplier with Receiving Voucher (RV)/Delivery Note (DO). If parts are for UL/CSA or BEAB models, Material Certificate/Packing List must be attached.

2) Store will sign the RV/DO after confirming the materials (parts numbers) and quantities. Store will then despatch the RV/DO as well as Material Certificate to Incoming QA Section for inspection.

3) Incoming QA will differentiate which goods need to be inspected based on the criteria of past rejected rate.

4) For non-inspection goods, RV/DO will be signed and returned to store.

5) Listed parts (UL/CSA or BEAB) must have Material Certificates attached. If this is not so, the lot is considered to be rejected.

6) Inspected goods will be sampling checked by Incoming QA according to Single Sampling Plan for Normal Inspection (Level II) Military Standard-105D.

    a) Plastic :- Using AQL level 1.5%

    b) Mechanical parts :- Using AQL level 1.5%

    c) Electrical parts :- Using AQL level 1.0%

    d) Wire parts :- Using AQL level 1.5%

    e) Critical indications :- Using AQL level 0%

    f) Others :- Using AQL level 2.5%

7) All inspection parts must be checked according to their individual inspection manual.

8) For inspection passed lots, RV/DO will be signed and returned to store. At the same time, RV/DO are despatched to Material, Purchasing and Account sections. Account section in turn despatches RV/DO to our company's Delivery section and Supplier's Account section. In conjunction with this, rejected parts found during inspection are returned to store for claiming purposes. Supplier's QC will be informed about the defective symptoms through Quality Feedback Form if deemed necessary.

The procedure state below is depicted in "FLOW CHART FOR INCOMING REJECTED GOODS REF. NO. ICAQ/021/85".

9) For rejected goods, Incoming QA will issue "Incoming Inspection Report" (QIM) which will be despatched to Material, Purchasing and Account sections. Account section in turn despatches QIM to Delivery section and supplier's Account section. At the same time, Incoming QA will inform supplier's QC section about the defective symptoms of the rejected lot, paste pink form on the goods and indicate "rejected samples" on carton boxes containing rejected samples.

10) If supplier's QC agrees with the rejected level, purchasing section will make arrangement for the supplier to rework in our company, take back for rework or replace the rejected lot with new lot.

11) If supplier QC does not agree with the rejected level, a discussion between supplier's QC and our company QA and/or Engrg. is carried out. After discussion, if the explanation given by supplier's QC is acceptable, 'Application for Special Acceptance' form is submitted to our company. Otherwise, procedure stated in Item 10 is followed.

12) For the rejected lot taken back by supplier, procedure starting from Item 1 is followed again.

13) For the rejected lot reworked by supplier in our company, our company QA will re-sample the lot. If sampling fails, level setting will be discussed between Supplier's QC and our company QA. However, if sampling fails three times, QA Manager is informed and a discussion with supplier is carried out.

14) If sampling passes, Purchasing section will issue "Rework Voucher for Rejected Material" (RM) to Incoming QA for completing.

15) After completing the RM, accepted goods are returned to store together with the RM. Store will sign the RM and return it to Incoming QA after confirming the quantity. The goods are then kept in store prior to issuing to production lines.

16) Incoming despatches RM to Material Control, Purchasing and Account sections. Account section in turn despatches RM to Delivery section and Supplier's Account section.

REF. NO.: ICQA/006/85		
PREPARED BY :	DATE :	25/5/85
CHECKED BY :	DATE :	25/5/85
APPROVED BY :	DATE :	25/5/85
REVISION :		
DONE BY :	DATE :	1/4/ 86
APPROVED BY :	DATE :	1/4/ 86

APPENDIX 9

STORAGE PROCEDURE

1. **STORAGE OF MATERIALS - Use of "FIFO" possible**

   "FIFO" means "First In First Out".

   (a) All Electronic components, whether Bulk or Taping form, i.e., RESISTOR, CAPACITOR, DIODE, IC, TRANSISTOR COIL, PCB, TRANSFORMER, WIRES and CONNECTORS.

   (b) BIG PARTS, i.e., CRT, B/C, F/P, FBT, D/Y, metal/plastic parts.

2. **RECEIVING SECTION**

   (a) Check quantity/parts specification, according to Supplier Delivery Voucher.

   (b) Listed Inspection goods (QC's Critical Components List), inform Incoming QC about inspection work.

   (c) If goods are accepted, Incoming QC must stamp "ACCEPTED DATE" on the box/roll/package containing the materials. For non-inspection goods, goods are to be verified by store personnel and stamped or marked with "RECEIVED DATE" on the box/roll/package containing the materials.

3. **PACKING SECTION**

    (a) Store personnel in-charge should prepare for packing those materials received in the earlier stage by noting the dates accepted or received on the packages containing the materials.

    (b) Proper packing could also be done by paying attention to the Purchase Order Number sequence, e.g., BT-2032P(I)-1, 2, 3 and so on.

4. **"FIFO" STORAGE SYSTEM**

    (a) For components, each has a coded compartment on the racking system. If the compartment is filled, the same components are to be placed on the reserved rack, until the components are used up by the packer.

    (b) For BIG PARTS such as CRT, B/C, F/P, FBT, D/Y, etc., these are stored according to either "RECEIVED DATE" or Purchase Order Number sequence.

    (c) For Metal/Plastic Parts, they will be based mainly on QC "ACCEPTED DATE" for packing. Therefore, in applying the "FIFO" system to such materials, the materials must be stored according to arrival dates at specific areas.

5. All Store personnel are to follow strictly the "FIFO" system for the smooth running of the Store section.

APPENDIX 10

FACTORY EQUIPMENT - CALIBRATION

Equipment To Be Calibrated At SISIR

1) Electrostatic Voltmeter
2) Distortion Analyzer
3) Spectrum Analyzer
4) LCR Meter
5) Q Meter (one unit in plant only)
6) X-Ray
7) Colour Analyzer (NUS?)
8) Thermometer (one unit in plant only)

CALIBRATING EQUIPMENT	EQUIPMENT TO CALIBRATE
DC Meter Calibrator	
Digital Multimeter	
AC Meter Calibrator	Ammeter & Voltmeter
High Voltage Generator	
AC High Voltage Meter	
High Frequency Voltmeter	
Frequency Counter	
Frequency Standard	Signal Generator
AM/FM Modulation Degree Meter	
Distortion Meter	

High Frequency Voltmeter
Frequency Counter
Frequency Standard — Oscillator Generator
Distortion Meter

Oscilloscope Calibrator — Oscilloscope

Standard Resistor
YEW 2755 Bridge — Insulation Tester

AC High Voltage Meter
High Voltage Generator — Hi-Pot Tester
Hi-Pot Calibrator

Frequency Counter
Frequency Standard — Frequency Counter

Luminance Meter — Luminance Meter (Calibration In HQ)

Gauss Meter
Standard Magnet — Gauss Meter

APPENDIX 11

MEMORANDUM

Date   : 17th. July 1986

From   : Engineering Manager

To   : All Engineering Members

RE: QCC

We have been requested to form a Quality Control Circle within the Engineering Section as soon as possible.

For your information, Quality Control Circles have to be formed/re-organised as follows:-

(a) Production Section - 4 QCCs

(b) Materials Section - 1 QCC

(c) QC Section - 1 QCC

(d) Engineering Section - 1 QCC

The objective of having a Quality Control Circle(s) within each department is to have a team, which is familiar with the department's own peculiar problems and which is in the best position to help improve the job efficiency and quality of the department and the working environment, to meet regularly and come up with possible solutions to problems which may crop up from time to time within the department, and to participate in all other QCC activities.

We have now to select a Group Leader and a group of say, ideally, 7 people (including the Group Leader) at the impending QCC meetings. Your suggestions are required and are most appreciated.

Please submit lists of QCC "discussion" topics, which you feel are important, to the undersigned by latest, <u>Monday, 21st. July `86, 10.30 am</u>.

Thank you.

Engineering Manager

cc

P/S:- Suggestions or volunteers, regarding the selection of the QCC members and their Leader, are most appreciated.

APPENDIX 12

TEST SIGNAL AND ITS APPLICATION

The test signal is symmetrical geometry; it is used to check the response on various frequency, phase shift, centering and also optimal adjustments.

1. **Type of Test Signals on T8**
    1.1 Circle pattern
    1.2 Cross hatch pattern
    1.3 White pattern
    1.4 Jumping signal

2. **Application**

    2.1 Circle pattern

    2.1.1 Cross hatch raster

    It provides checking of:

    (a) The picture geometry, such as horizontal and vertical scanning amplitude and linearity. The raster should form equal square.

    (b) The uniformity of focus and the pin-cushion correction. The focus of the cross hatch area and of the central area should be uniform, while at the same time the lines of the cross hatch should seem to be straight and parallel at normal viewing distance.

    2.1.2 Circle

    Diameter - 12 units of the cross hatch raster, the centre of the circle also being the centre of the picture. It provides checking of:

    (a) The picture geometry and scanning linearity.

    (b) The picture aspect ratio. The circle should appear truly circular if the picture has the standard 1.33 aspect ratio (4 : 3).

    2.1.3 Black rectangle in the top area of the circle

    It provides checking of the low-frequency response. Poor response shows as streaking from the edges of this rectangle to the right-hand side.

    2.1.4 White rectangle with black needle in the top area of the circle

    It provides a check on whether reflections are present in the received television signal.

2.1.5 Black bar with white crosses in the central area of the circle

(a) The horizontal white line is composed of two TV lines, one in each field. As their scanning rhythm is contrary to other horizontal white lines it provides a very effective checking of the interlacing. Any faulty interlacing appears as a deviating thickness of this horizontal white line, compared with the other horizontal white lines.

(b) The white cross indicates the centre of the picture.

2.1.6 Blocks of frequency gratings

Five blocks of gratings each consisting of vertical strips corresponding to the following frequencies = 0.8 - 1.8 - 2.8 - 3.8 - 4.8 MHz. They provide checking of resolution and bandwidth. The gratings are sine-wave signals and should (possibly with the exception of the 4.8 MHz block) appear.

2.1.7 Greyscale steps

Six rectangles of 0% - 20% - 40% - 60% - 80% - 100% video amplitude. They provide checking of the linearity of the transmission path and adjustment of correct brightness and contrast of the picture. That is, the adjacent rectangles should show a constant change in contrast.

2.2 Cross hatch pattern

This pattern is used for adjusting the convergence of CTV receiver. Its line frequency is 15625 Hz ± 1 Hz and frame frequency is 50 Hz ± 1 Hz. Cross hatch is 15 x 15.

2.3 White pattern

White pattern is a pattern with blank raster (white), it can be easily utlised to check the fault of cross-over distortion and ringing effect. Two vertical black lines (indicated by the letter 'a') is for checking the horizontal sync.

2.4 Jumping signal

Jumping signal in video checker board pattern is used to check the automatic synchronization of monochrome TV sets.

Line frequency - 15625 Hz + 2% and 15625 Hz - 2%

Cycle time - 15625 Hz + 2% and 15625 Hz - 2%
during 3.5 sec. each with intervals of 0.5 sec.. Complete cycle time is 8 seconds.

www.ingramcontent.com/pod-product-compliance
Lightning Source LLC
Chambersburg PA
CBHW081303170526

45165CB00011B/3393